"Share your knowledge. It is a way to achieve immortality." 14th Dalai Lama

SUSAN'S OTHER SPIRITUAL BOOKS

Treasures from the Spirit World

Seers, Mediums and Spiritualists

Spiritualist Camps: Their History and Prominent People

Blessing from Agnes

THE HEAVEN HYPOTHESIS

SUSAN URBANEK LINVILLE, PH.D.

Pokeberry Press

The Heaven Hypothesis

Published by Pokeberry Press.

ISBN: 978-1-951847-07-4

Printed in the United States of America

Cover and Book Design -Susan Urbanek Linville and Stephen V. Ramey

Cover art: *Gemini* (2026) Heaven Hypothesis Cover, https://gemini.google.com

AI-generated image.

Contents

Introduction

What happens when a freelance writer with a PhD in biology becomes friends with a near-death experiencer/medium who has her own Spiritualist church? You end up with a book called The Heaven Hypothesis.

My interest in science goes back as far as I can remember. One of my earliest memories was sitting in the living room with my dad watching astronauts shoot off into space. On February 20, 1962, just after my 6th birthday, John Glenn flew the Friendship 7 mission, becoming the first American to orbit the Earth. What could be more exciting than that? I wanted to be an astronaut. Over the next decades, I watched every launch: Mercury, Gemini, Apollo, the space shuttles. I even saved newspaper articles.

When I was about eight, my family visited the Cleveland Museum of Natural History. The museum's new building is recognized as one of the most beautiful in the world today. At the time, I could have cared less about the building. What enthralled me were all the fossils, specifically dinosaur fossils. By the time I was in 5th grade I had a dinosaur diorama set up in my bedroom with little plastic dinosaurs living in a paper jungle. I even had a shallow dish filled with water to create a sea for my aquatic animals. I wanted to be a paleontologist. My mother probably thought she'd brought the wrong baby home from the hospital.

By the time I reached adulthood, even though I was interested in all kinds of science, I narrowed down my interest to a degree in Biology. For my PhD, I studied bird behavior. Specifically: *Sexual Selection and Plumage Ornamentation in a Socially Monogamous Passerine, the*

Northern Cardinal (Cardinalis cardinalis). I spent hours in the woods, trapping birds, banding birds, taking blood samples, measuring plumage color in males and females, learning DNA fingerprint techniques in the mid-90s (much more difficult back in those days), locating nests, and recording feedings. After three years gathering data and going through all the statistical analysis, I found that brighter red males were better fathers than duller males (feeding the nestlings more). It wasn't as exciting as being shot into space atop a controlled explosion, but it was science, and I loved being outside. Of course, we know now that birds have descended from dinosaurs, so I guess you could say I was studying dinosaurs.

I didn't follow the normal course of an academic. Trying to balance divorce and remarriage, single motherhood and family, I ended up teaching part-time and then writing about science. I discovered that translating research from scientific journals into a story that the public could understand was just as important as teaching what an endoplasmic reticulum was to pre-nursing students. I spent years writing hundreds of podcasts for Indiana University's "A Moment of Science."

My life wasn't all science. Many trips to the library introduced me to Egyptians with their Book of the Dead, and other ancient cultures. Then there were the more esoteric books about the Lost City of Atlantis, ancient astronauts, and all the topics that made *Raiders of the Lost Ark* a hit movie. Edgar Cayce's life story and predictions managed to drift into my life several times as well as meditation and eastern religions. I've been a Catholic, Presbyterian, a member of the church of Unity, and lastly, an observer of Spiritualism.

In September of 2012, friends invited me to a Murder Mystery Dinner to benefit a small congregation at the Spiritual Path Church in New Castle, Pennsylvania. I had read plenty about meditation and psychic phenomena since I was a teen, but I didn't know anything about Spiritualism. At first, I didn't even connect the church to seances and communicating with the dead.

I met Reverend Karen Heasley and her friend and assistant minister, Marilyn Fair, at the ticket table when we entered the dining hall. They seemed very welcoming, but the place was packed, so I did little more than say hello. The meal was a typical western Pennsylvania dinner with chicken, pasta, salad and dessert. The actors were from the local theater and they made the night a memorable one. The crowd eventually solved the mystery. We identified the murderer. Little did I know I would soon return to the church.

At the time, I was working at the local historical society, researching the history of some of the city's Victorian homes. The society has a large collection of photographs from the local area on file. A couple of days after the dinner, out of the blue, I discovered there were two photographs of members of an old Spiritualist church in the collection. One appeared to memorialize an important event. Seven people posed around a table which held a certificate. At the base of the photo in white hand-lettering it said: Spiritual Church of Truth, July 18, 1921. I thought the photo might be of interest to Karen. It was.

One thing led to another, and I was soon writing blogs about Spiritualism to post on the church's website. For nearly a decade, I worked for Karen, researching the history, the people, and the influence of Spiritualism in the United States and around the world. That led to hundreds of blogs and four self-published reference books on the topic.

I soon discovered that Spiritualism was not what is portrayed in TV and movies. Mysterious mediums do not appear from the shadows. Crystal balls, Ouija boards and Tarot cards are unnecessary. There is no fortune telling or palmistry. To define Spiritualism, one must look back to one of its early proponents, Emma Hardinge Britten. Britten was born in England in 1823 and was an influential leader in establishing early Spiritualism. She was a passionate orator and writer on the spiritual movement, as well as a practitioner. She published two significant books, *Modern American Spiritualism* (1870) and *Nineteenth Century Miracles* (1884), which are detailed accounts of the early history of the modern spiritualism movement in America.

According to Britten, Spiritualism is "the discovery of a mode whereby discarnate spirits can and do communicate systematically with earth." Spiritualism teaches that people are, first and foremost, spirits. The body is a material vessel in which the spirit grows and becomes an individual. Once the body dies, the spirit continues eternally.

Communication is the foundation of Spiritualism. Spiritualists are not satisfied to rely on faith alone, as many other religions. The medium's goal is to demonstrate the existence of a spirit world by contacting spirits who have passed from the earthly realm. This means identifying the person who has passed and determining that the communicants are the spirits of the men, women and children who once lived on earth and are now in the spirit world.

I also discovered that Spiritualists were serious about mediumship and the accuracy of their communications. I'm not saying that there were no impostors, quacks and phonies. Like today, there were plenty of people ready to take advantage of the public. There were also many scholars trying to test the genuineness of mediums. James H. Hyslop was one of the prominent researchers at the time who tried to verify the abilities of mediums. He wrote several textbooks and became involved with psychical research while at Columbia University. He founded the American Institute for Scientific Research in 1904 and took on the role of chief investigator with Hereward Carrington and Walter F. Prince assisting. They printed their first *Journal* in January 1907. Hyslop published several books on psychical research, including: *Science and a Future Life* (1905), *Enigmas of Psychic Research* (1906), *Borderland of Psychical Research* (1906), *Psychical Research and Survival* (1913), *Life After Death* (1918), and *Contact with the Other World* (1919).

When I began to mull over the idea of a book about heaven and life after death, I found there were plenty of books and articles out there. Religious books on heaven, usually Christian. Near-death experiencers and their view of heaven. Religions of various ancient societies. Consciousness and brain science. Psychology and hallucinations. Burial practices of ancient humans, and even animal burials. Heaven and science are placed neatly into different categories. Except by a few medical doctors studying near-death experiences, researchers tend to avoid the topic of heaven or anything spiritual.

I wondered if there was a way to meld all these concepts together, to look at prehistory, history, religion, brain science, and near-death experiences. Early in the 20th century, Spiritualist researchers tried to prove that the spirit lived on after the body perished with what little technology they had. They weren't trained in hypothesis testing but tried to be rigorous with their techniques. Today we not only have the technology to examine the inner working of the brain, but we can also meticulously test for the existence of heaven.

This book is my attempt to blend the scientific and the spiritual, to examine humankind's history with heaven and the afterlife, and to review the research that is happening today. Can we prove that the consciousness can transcend the body? Is there a place, a dimension, or space-time where spirits exist? I can't answer that, but I do know we can now detect subatomic particles unheard of a century ago and see the light from galaxies that formed near the beginning to the Big Bang. I have confidence that with some ingenuity we can also test the Heaven Hypothesis.

SECTION I
THINK LIKE A SCIENTIST

Chapter 1
What is a Hypothesis?

Before we talk about developing a Heaven Hypothesis, it is important to understand what a hypothesis is. A scientific hypothesis is the initial building block of the scientific method. It is an "educated guess" about what causes something to happen based on prior knowledge and observation. Let us consider an example. What if you notice that the plants in your house grow better in rooms where they are exposed to music? Plants don't have ears to listen to music, but what if the sound vibrations affect their growth in some way? You could just say that all plants like music because you've noticed a size difference in your house. But how do you know music is affecting the plants? Maybe it's the sunlight or temperature of the rooms, or even the humidity.

Your educated guess needs to be tested in a non-biased way. And scientists have done just that, tested to see if sound affects plant growth. It appears that both plants and germinating seeds respond to sound. One research group found that certain sounds could improve growth in cucumbers, peppers and tomatoes. Others found that okra and zucchini seeds exposed to natural sounds sprouted better. Certain sounds could also help rice paddy seed germination, plant height, and root system development.

Hypothesis testing has only been around for about 450 years. It all began with the famous astronomer Galileo Galilei. Galileo was born in Pisa, Italy, home to the Leaning Tower, in 1564. The city is located along the Arno River, about 6 miles from the Ligurian Sea. It was a chief port in Galileo's time, with spectacular cathedrals, palaces, and villas.

Galileo was the oldest of six children born to Vincenzo Galilei. Vincenzo was both a musician and music theorist. When the family moved to Florence in 1574, Galileo was fortunate to receive one of the best educations of the time. He attended Camaldolese Monastery and entered the University of Pisa to study medicine, but fell in love with mathematics and physics instead. After graduating from the university, he supported himself with small teaching jobs while studying mathematics and objects in motion.

Let's go back in time a bit. Centuries before Galileo was born, Greek philosopher and naturalist, Aristotle, studied everything but the kitchen sink. He wrote books on a variety of subjects including physics, astronomy, geology, biology, psychology, and philosophy. Like Galileo, one of his interests included the motion of objects. He claimed that, if dropped, heavy objects would seek "their natural place" faster than light ones. His hypothesis was that heavy (massive) objects fall faster than lighter ones. It seemed like a logical prediction. Of course, in those days, there was no hypothesis testing. No one questioned Aristotle's statement. He said it was so. It must be true.

When he was 26, Galileo was working as a math teacher at the University of Pisa. He realized that Aristotle had made an educated guess about falling objects. But what if Aristotle's guess wasn't correct? Galileo decided to find out.

To test the hypothesis that a heavy object would fall faster than a lighter one, Galileo dropped two balls of different weights (mass) from a high tower. He found that the heavy ball hit the ground first, but only by a small fraction of time. The lighter ball always started out a bit faster than the heavy ball. Then the heavy ball caught up. Except for a small difference in air resistance, both balls reached the ground at nearly the same speed.

It seemed crazy. Why would objects of different weights fall at the same rate? Galileo couldn't explain it. Had he proved Aristotle's assumption wrong? He tried another test just to be sure. Using an inclined plane this time, his experiment slowed the acceleration of the balls enough so that the elapsed time could be measured more accurately. Balls were released to roll a known distance down the ramp. The time it took for each ball to move along that distance was measured using a water clock. Again, Galileo's experiments showed that heavy and light objects fell at the same rate.

Galileo was a true rebel of his time, both in his personal life and professional one. Unfortunately, Galileo's criticism of the famous Aristotle was not taken well. In 1592, he was fired from his position at the University of Pisa. After finding another job at the University of Padua, he met Marina Gamba, a woman who lived with him out of wedlock (perhaps for financial reasons) for years. They had three children together, which caused another uproar. His children were illegitimate. The girls joined convents because of their low social standing. Only his son's birth was legitimized, and he became a successful musician.

Galileo's research didn't stop with the motion of objects. He delved into astronomy, using some of the earliest telescopes. He supported Copernicus' view that the sun was the center of the solar system, which got him reported to the Roman inquisition. He was found "vehemently suspect of heresy" by the Catholic church in 1633 and sentenced to house arrest for the rest of his life.

Fortunately, things have changed dramatically since Galileo's time. Hypothesis testing is used today for basic scientific research in all fields from astronomy to zoology.

The story of ulcers is a good example of how hypothesis testing has helped millions of people. Until the 1980s, doctors believed all ulcers were caused by stress and eating acidic foods. They recommended a diet low in spices, assuming those foods were the root cause of the problem. In 1982, Australians, Dr. Barry Marshall and his colleague, Robin Warren, noticed that people with ulcers had certain bacteria called *Helicobacter pylori* present in their digestive tracts. Maybe stress and acid didn't cause ulcers. Maybe it was something else, a bacterial infection. But how could Marshall know for sure if the bacteria were responsible?

First, he needed a hypothesis. Marshall's hypothesis was that *H. pylori* caused ulcers. He needed to test that hypothesis to see if his guess was correct. Scientists don't usually experiment on themselves, but that summer, fellow scientist, Neil Noakes, took bacteria from a petri dish, mixed them with lukewarm beef and handed the mix to Marshall to drink. After downing the cocktail, Marshall suffered severe inflammation of the stomach. A link between an infection with *H. pylori* and the development of peptic ulcer disease looked like a possibility. But Marshall was just one person reacting to the bacteria. This single trial didn't prove or disprove his hypothesis.

In 1984, Marshall and Warren obtained funding from the Australian Medical Research Council to conduct an official double-blind research study of *H. pylori*. They used 100 patients to determine if antibiotics could cure duodenal ulcers and found that the antibiotics worked. Marshall and Warren received the Nobel Prize in 2005 for their groundbreaking research.

Hypothesis testing is also used to ensure that the pharmaceutical industry is selling safe and effective drugs. In the 19th century, anything could be sold as medicine. An advertisement for Wine of Cardui, for example, claimed that “more than 500,000 American women have taken McElree’s Wine of Cardui during the past year. These women were afflicted with every kind of ‘female trouble.’” They advertised that “ninety-nine out of every one hundred of them are completely cured....” Was this statement true? There were no regulations to protect the consumer. Women had no idea what was in it. As a matter of fact, the main ingredients of this magical tonic were potassium (51.9 percent), salt (16 percent) and a varying amount of alcohol up to 23.3 percent. No wonder they were feeling better- they were drunk.

Because hundreds of products made false or dubious claims, Congress passed the Food and Drugs Act of 1906. It prohibited drug manufacturers from using false or misleading drug labels. Many drug companies went out of business or changed their labels from "cure" to "remedy" to escape the regulations. It also required that any dangerous or addictive ingredients, such as opium, heroin, cocaine and alcohol, be listed on the label. One Night Cough Syrup’s label of ingredients for example included alcohol, cannabis, chloroform, and morphine. Doctors also used morphine to relieve many female patients’ menstrual cramps and morning sickness. By the late 1800s, women accounted for more than 60 percent of opium addicts.

Today, every drug must be evaluated. Does Drug A lower blood pressure? Does Drug B reduce pain levels? We know that when researchers test drugs there can be a placebo effect. This effect is a phenomenon where a person’s condition improves after they receive a fake treatment like a pill with no drug in it. Hypothesis testing is needed to show that drugs work and aren't just a placebo response.

Hypothesis testing is used in basic science as well as in medicine and the pharmaceutical industry. The question is, can the idea that heaven exists undergo hypothesis testing? The trending thought today is that science cannot test the supernatural. It can only test the physical world. Mystical phenomena fall outside the scope of scientific research. But the boundary between the natural and supernatural has been shifting over time. Disease, lightning, and heavenly objects like comets were once thought to be supernatural. Now they are considered part of the physical world.

We now know the natural world is stranger than Galileo could have imagined. The universe is made of subatomic particles that live a quantum existence, blinking in and out of reality. For instance, light behaves as both a wave and a packet of light called a photon. While light is traveling, it is moving as a wave. When it interacts with something, like an experimental measuring machine, it becomes a photon with specific characteristics. One characteristic is known as spin. While it is a wave, it can have any spin. There is no way to know the spin until it is measured.

One of the strangest characteristics of quantum physics is entanglement. That can happen when two photons are emitted during an experiment at the same time. In 1972, Stuart Freedman and John Clauser found if they put a bunch of calcium atoms into a highly excited energy level under certain conditions, the atoms emit *two* photons. Even though the photons are separate and traveling off in different directions, if the researcher measures one photon and it has a left spin, the other one will have the opposite right spin. Always. The photons could be 100,000,000 miles apart and the results would be the same.

In 1947, Einstein said this about entanglement: "I cannot seriously believe in it because the theory cannot be reconciled with the idea that physics should represent a reality in time and space, free from spooky action at a distance."

Einstein might have thought it impossible, but research has shown us that two photons can be entangled. Since Freedman and Clauser's first experiments, entanglement has been observed with atoms and electrons. Physicists have recently entangled 15 trillion electrically charged atoms in a hot soup at the National Institute of Standards and Technology. They hope to use the new process to build hybrid quantum computers that can manipulate, store, and transmit different forms of data.

If the quantum world wasn't strange enough, we now suspect there might be matter and energy in the universe that we can't detect or measure with any equipment we have today. They are called Dark Matter and Dark Energy. Unlike normal matter, dark matter does not absorb, reflect, or emit light, making it extremely hard to find. Scientists have only been able to infer the existence of dark matter from the gravitational effect it exerts on visible matter. And dark matter seems to outweigh visible matter six to one, making up about 27% of the universe. Dark energy appears to be distributed evenly through the vacuum in space. Scientists think it accounts for a repulsive force, which tends to accelerate the expansion of the universe, but it remains a mysterious substance.

The natural world seems to be more bizarre and inexplicable than we ever imagined. Physicists are pushing the boundaries of the natural world into the study of the multiple dimensions of String Theory. Some are even suggesting that the universe could be a projection from an outside source. At the same time, biologists are trying to define consciousness. And studies of Near-Death Experiences are giving us accounts of tantalizing instances of consciousness existing beyond the physical brain.

If we are going to test for heaven we must deliberate like scientists. That's where critical thinking comes into play. What do we need to test the hypothesis? What are our existing biases about the subject? Some say heaven exists because the Bible says so, others say it's just a myth. Is there some middle ground to work from? I think that there is.

Chapter 2
Critical Thinking

If we want to develop a hypothesis about heaven, we need to discuss critical thinking. The whole idea of critical thinking began with the ancient Greeks. Socrates is given credit for proposing the idea. He was born as an Athenian citizen about 470 BC to a relatively affluent stoneworker. Like most wealthy Athenians, he was well educated, learning music, poetry and gymnastics as well as basic reading and writing. He also fulfilled his military service, fighting in three campaigns during the Peloponnesian War.

Socrates lived close to his father's family and inherited part of his father's estate which gave him the free time to develop his ideas about philosophy. He was a great teacher and orator, but he never wrote down his philosophical ideas. It was Socrates' student, Plato, who memorialized Socrates' ideas through his writings. According to Plato, Socrates stated simply that nothing should be accepted out of hand. One must question ideas through the analyses of beliefs, assumptions, and presumptions, and by relying upon evidence and sound rationale.

The U.S. National Council for Excellence in Critical Thinking in 1987 defined critical thinking as the "intellectually disciplined process of actively and skillfully conceptualizing, applying, analyzing, synthesizing, or evaluating information gathered from, or generated by, observation, experience, reflection, reasoning, or communication, as a guide to belief and action."

In other words, critical thinking involves considering different points of view and weighing evidence carefully. The world is not black and white. To think critically, one needs to see the world in a more nuanced way and understand how things are connected. It's important to be skeptical, think objectively, and be able to spot connections between ideas. A critical thinker challenges assumptions, recognizes their own biases, and closely assesses the validity of arguments and solutions.

There are several steps involved with critical thinking:

1. Identify the topic to clearly define the issue or question at hand to focus on the right issue.

2. Collect relevant information or data from reliable sources, including preexisting research and experts.

3. Evaluate the information for accuracy, reliability and relevance.

4. Consider different viewpoints to reduce bias in one's decision making.

5. Consider possible outcomes based on the analysis.

6. Choose the best course of action.

7. Reflect on the outcome of the decision to improve future critical analysis.

We would all like to think that we are able to judge what information is true and what is false. Unfortunately, critical thinking is not something we learn in school. We are all subject to what is called

Conformation Bias. This bias is a person's tendency to accept and look for information that is consistent with their existing beliefs. Because we are bombarded with information constantly, particularly by corporations and political planners, it is almost impossible to critically analyze everything. On top of that, internet algorithms amplify our own biases by limiting what information we are exposed to, usually for the benefit of the internet provider.

Scientists believe that people are susceptible to confirmation bias because it is an efficient way to process the information that they are bombarded with each day. We cannot take the time to carefully process each piece of information to form an unbiased conclusion. The need to process information quickly to protect ourselves from harm has led to a biological system that can make errors. It can also be a way to protect one's self-esteem. Discovering that a highly held belief is incorrect makes people feel bad about themselves. Therefore, they tend to seek information that supports their existing beliefs.

Research has shown that confirmation bias is strong and widespread. Information that conflicts with a person's decision may cause discomfort, and the person will therefore ignore it or give it little consideration. It is not that people aren't incapable of accepting arguments that are counter to their beliefs, but, rather, people aren't motivated to do so.

I'm sure entire books have been written about confirmation bias, so we will not cover it in detail. The important thing to understand is that confirmation bias is largely unintentional, and we all do it, even scientists. That is why hypothesis testing is so important. It is a way to eliminate or at least reduce the biases that we all have.

Let's look at the example of Confirmation Bias. In 1968, the Swiss hotel manager, Erich von Daniken, published a book that was entitled,

Chariots of the Gods in English. He claimed to have evidence that aliens from outer space had been important in shaping human history. The idea was so intriguing to the public that his paperback sold over 7 million copies. By 1974, he had sold his second book, *Gods from Outerspace*, and was convinced that aliens had altered human intelligence through genetic engineering.

Von Daniken's "proof" for his hypothesis came from his interpretation of many archeological sites and artifacts from around the world. If you examine his books closely, you will notice that he uses "argument by assertion" instead of critical analysis. These arguments fail three steps of critical thinking. First, he gathers information, but only the information that backs his preconceived idea (confirmation bias) that aliens have influenced human development. Second, he analyses the information in a biased way, not taking all information into consideration. Third, he doesn't consider different perspectives from experts who work in the field.

For example, he proposed that a sarcophagus carving in Palenque, a late-classic Maya site, in Chiapas, Mexico, was an alien flying a spaceship. He doesn't consider the context in which the stone was carved or anything about the beliefs of the people. He sees a "spaceman," while trained archeologists who have been studying the culture know that the scene depicts, Pakal falling backward into the underworld. The spaceship is a stylized depiction of the World Tree of Maya cosmology.

In another case, he claimed that the Delhi Pillar, a column of iron that has not rusted in 4,000 years, couldn't be explained. Of course, he concluded it was created by aliens. It's not nearly as old as he declared, being created in the 4th century. When metallurgists analyzed the pillar, they found it to be 99.72 % pure iron and 0.1 % phosphorus. The lack of rusting was due to several factors: the presence of phosphorous, the dry Delhi climate, a thin layer of scale produced when it was forged, and ceremonial ghee (clarified butter) that was applied to it for over 500 years. The mystery was explained by simple chemistry, not by including aliens.

Thinking that aliens visited ancient people might seem like harmless entertainment. What's the problem with readers believing what von Daniken wrote in his books? Everyone loves a mystery, including me, be it Aliens, or Big Foot, or the Lost City of Atlantis. Maybe some of von Daniken's ideas were true. The problem is, they were never critically analyzed by the author.

Von Daniken's ideas may have been unfounded, but his books lead to seriously detrimental effects on people, their families, their health, and their lives. But other ideas have. When presented by someone with authority or claims of expertise, people can be misled. An "argument by assertion" can be used to promote ideas that are distorted and even life threatening.

One example of a dangerous "argument of assertion" is the alleged connection between vaccinations (specifically the mumps, measles, rubella, MMR vaccine) and autism. People who claim to have medical authority argue that increasing vaccination numbers have risen with an increase in autism. They conclude that the two events are linked, but it takes lots of scientific evidence to show that one condition causes the other. Has anyone shown definitive proof that the two are connected?

It appears they have not. According to the World Health Organization, "Evidence based on 31 primary research studies, published between January 2010 and August 2025, including data from multiple countries, strongly supports the positive safety profile of vaccines used during childhood and pregnancy, and confirms the absence of a causal link with ASD (autism spectrum disorders)." Recorded autism rates have indeed risen substantially in recent decades, but that doesn't mean that autism is occurring more often. Instead, the rise reflects a change in the definition of autism, creating a spectrum of conditions being included under one umbrella.

The vaccine-autism link was first proposed in the 2000s by medical researcher Andrew Wakefield. Wakefield graduated from St. Mary's Hospital Medical School in 1981 and studied transplantation at the University of Toronto in the late 1980s. In the 1990s he joined the Royal Free Hospital in London where he worked on inflammatory bowel disease. In 1998, he moved to Austin Texas where he co-founded Thoughtful House which studied autism and promoted he's anti-vaccine ideas. He appeared to have the medical credentials to make a claim about vaccines and autism, but did he have the evidence to back up his claim or was he using "argument by assertion?"

Wakefield's honesty was first put to question when he published research in 1998 in the respected medical journal, *The Lancet*, involving 12 children with bowel inflammation that occurred after receiving the MMR vaccine. He then linked the inflammation to autism.

The good thing about hypothesis testing is that experiments and results should be able to be replicated by other researchers. In this case, other researchers couldn't validate his claims. Soon after, a newspaper reporter discovered that Wakefield stood to earn millions of dollars selling test kits for the inflammatory condition. After a two-year investigation, the British Medical Council concluded that he had engaged in unethical and dishonest research. *The Lancet* retracted Wakefield's research paper in February 2010, saying he had falsified elements of the manuscript. He resigned as executive director of Thoughtful House.

Since Wakefield's research claims were widely reported in the press at the time of publication, people worried that vaccines caused autism. It was better to be safe than sorry. There was a sharp drop in vaccination rates in the UK and Ireland, and then the United States. Although the claim was refuted, anti-vaccination propagandists didn't want to disrupt their biased opinions. For the next three decades, their continued support of the idea led to an increase in the incidence of preventable childhood diseases, resulting in deaths and permanent injuries.

Measles is one of those diseases. It is a highly contagious virus, and often comes with high fever, cough, runny nose, watery eyes, rash and Koplik spots. Complications can include diarrhea, ear infections, pneumonia, encephalitis (brain inflammation) and subacute sclerosing panencephalitis (a rare but fatal brain disorder).

In 1924, there were over 511,000 cases of measles reported in the United States. With a death rate of 1-2 per 1000, hundreds of children were dying per year. The first vaccine was developed and distributed in the 1960s. By 1999, cases of measles per year in the United States had dropped to under 100. Childhood immunization was estimated to have saved 154 million lives.

Along with other diseases like polio, mumps, chicken pox and others, vaccines had become a success story. Unfortunately, anti-vaccine promotion during the past two decades has resulted in fewer immunizations being given. That was followed by an uptick in measles cases, most recently 2,267 in 2025 with two deaths. Outbreaks continued into 2026. Pertussis (whooping cough) is also on the rise, with only 3,962 cases in 2023 and more than 17,500 in 2024.

While the public is centered on the vaccine-autism connection, they are ignoring what may be the true cause of the condition. Over the last two decades, genetic studies have identified more than 100 genes harboring rare mutations linked to autism. Today, researchers are trying to discover how and why so many genes can cause the common features found in the condition.

Dr. Daniel Geschwind, the Gordon and Virginia MacDonald distinguished professor of Human Genetics, Neurology and Psychiatry at UCLA is searching for common biological changes that lead to autism. If he and his team can discover the molecular changes that occur during early fetal brain development, a critical period when autism-risk genes are most active, they may be able to identify and treat autism before a baby is even born.

Critical Thinking is important to both scientists and the public in general. Sometimes it can be a matter of life or death. For this book, I am going follow the premises of critical thinking the best that I can. I will first clearly define heaven. I've collected and evaluated relevant information from reliable sources, including preexisting research and experts. I've included multiple viewpoints to reduce bias. I make no conclusions, but I've suggested what I believe to be the best courses of action to test for the existence of heaven.

Chapter 3

Let's Define Heaven

Before we develop a hypothesis, we must first define heaven. Is it a supernatural place? Is it a location in another universe or dimension? Is it a state of mind or maybe a collection of consciousness in space and/or time? Let's look at some of the options that we will discuss in more detail in this book.

Heaven as a Supernatural Place

Heaven is defined in the dictionary as "a place regarded in various religions as the abode of God (or the gods) and the angels, and of the good after death, often traditionally depicted as being above the sky."

In the Abrahamic (Judaism, Christianity, and Islam) religions, heaven is considered the dwelling place of God. "The LORD has established His throne in heaven, and His kingdom rules over all" (Psalm 103:19). Heaven also seems to encompass the universe. "When I consider Your heavens, the work of Your fingers, the moon and the stars, which You have ordained" (Psalm 8:3). In Islam, Heaven is understood to be a place of everlasting life. According to the *Quran*, "But those who have faith and work righteousness, they are companions of the garden. Therein shall they abide forever" (2:82).

The Buddha did not believe in the existence of a supreme entity. He did teach about the existence of other worlds. These were heavens and hells populated by celestial beings. In some of the literature, Buddha is said to have gone to the heavens where he met with gods. Gods also descended to the earth to witness some events in the Buddha's life.

According to the Hindu Puranas, the universe is composed of fourteen worlds. The seven upper worlds are more heavenly, while the lower seven are places for wicked souls. *Bhuh* is the earth where we live. *Swah* is the world where people go after death to enjoy the rewards of their righteous actions. In the higher heavens, souls enjoy a spiritual communion with the personal God and cycle back to earth again. In the lower worlds, souls reap the results of their unrighteous actions on earth. The afterlife is composed of different worlds, bound by time, space, and causality.

For the ancient Egyptians, heaven was represented as a divine cow on whose back the sun god elevated from the earth, or as the goddess Nut arching over the earth. The afterlife could be in the fertile Field of Reeds, as a passenger in the solar ark, or among the circumpolar stars. In the *Book of the Dead,* all souls are judged by Osiris in the underworld.

In the 19th-century, the Spiritualist movement used mediums to communicate with spirits who had passed on to heaven. Some of those spirits described the new world through trance mediums. In *The Blue Island,* W. A. Stead described heaven as a set of spheres where souls worked toward self-improvement, moving from one sphere to the next. Communication was achieved through thought, and the passage of time was not the same as on earth. It is not clear if the location of the spirit world was a supernatural or physical location.

Heaven as a Physical Location

It wasn't until the 19th century that people began suggesting heaven was in a physical place. Joseph Smith, who founded the Church of Jesus Christ of Latter-Day Saints (the Mormons), claimed he experienced a series of visions in the early 1820s. He wrote that a heavenly body known as Kolob was "nearest unto the throne of God," which suggests that there is a heavenly realm in outer space. Physician and philosopher, D. Mortimer, located heaven within the sun in 1869. He described it as "a vast globe or world far within the surrounding photosphere of ethereal fire, the sun." Dr. Mortimer said that it was a divine revelation that convinced him "that the globe thus discerned is the Heavenly Empire wherein the righteous from this earth find their future home."

We haven't found any evidence of heaven in the sun or in nearby outer space, but what if heaven is another dimension? Our understanding of the physical universe is getting stranger all the time. To us, a table sitting in a room is a solid object. You can touch it, set stuff on top of it, and move it across the room. It is difficult to believe that the table is made of small particles that only have a certain probability of existing. Quantum Theory says that subatomic particles exist as fields and wink in and out of existence as physical particles, depending on their interactions with other particles. There is a probability, albeit a very miniscule one, that the table could vanish.

Another strange concept is String Theory. The mathematics of String theory, in an effort to explain quantum theory, says that physical matter is made of vibrating strings. The vibration determines how each string behaves in the macroscopic world. It also predicts other dimensions that could be curled in on themselves. Instead of the four dimensions we are familiar with, length, width, depth and time, these extra dimensions are hidden from us, maybe because they're tiny or warped. They might be invisible, but they can affect what we experience in the universe. Could heaven's location be wrapped in a string? At this point in time, string theory isn't complete and is difficult to test.

Some physicists believe an extra dimension might exist close to our own. Even though we cannot see it, we might be able to detect it by the amount of gravity it is sapping from our own dimension. The presence of another dimension might explain why particles have the masses that they do, which are less than what physicists predict. It might also explain why gravity is so much weaker than the other elementary forces in the universe. Maybe heaven is located there.

Heaven as a State of Consciousness

What is consciousness? The Greek philosopher, Plotinus, said that consciousness is composed of layers of experience that occur in different levels of the self. Higher modes of consciousness are layered over lower ones, and they enable human beings to draw on their own experience to investigate themselves and the nature of reality. Much later, Freud proposed that the mind was made up of the ego, id, and superego.

The study of behaviorism considered the brain to be a machine that responded to external stimulus. LeDoux examined neuroanatomy. Since humans have a highly developed frontal cortex, he suggested that that part of the brain might relate to knowing "what's on one's mind." Modern research is trying to isolate consciousness in the brain, looking for the minimal number of neurons needed to create a conscious experience. We are just now able to test some of these ideas with brain scanning machines.

Contemporary researchers have shown they can use a brain-scanning technique known as functional Magnetic Resonance Imaging (fMRI) to detect thought processes by indirectly measuring blood flow in the brain. Conscious awareness is associated with the cortex and a posterior area of the brain that includes the parietal, occipital and temporal lobes. But is conscious awareness equal to consciousness?

Eastern beliefs teach that the brain is an organ in the human body that it is a physical object. The mind has no location and is purely a mental phenomenon to be studied. Yoga and Vendanta call "pure consciousness" what remains when everything a person is conscious of has been removed. Buddhism teaches that consciousness is what exists after the mind is empty of everything that can be imagined. Those who practice meditation can reach different states of consciousness. Modern research on meditation has found that metabolic activity and respiration can be controlled by altering thoughts. The mind-body connection is just beginning to be revealed. Understanding consciousness is still in its infancy.

Near Death Experiences (NDEs) have been recorded since the Middle Ages. During these episodes, the subject is dead for a certain length of time. The brain is not functioning, but the person is aware of what's happening in the room where they are being treated. They may "travel" to another reality where they experience a variety of things including meeting with deceased loved ones, seeing vivid colors, and experiencing a feeling of peace and euphoria. Is this a hallucination of the dying brain, or is something else happening? Maybe heaven isn't a place, but instead the survival of consciousness after the death of the body.

If consciousness can transcend the function of the physical brain, can it be measured? In the 1970s, Raymond Moody documented hundreds of experiences from people who experienced NDEs. He found similar patterns in experience, but despite all the accounts, there was little evidence to prove that the experiences were not similar hallucinations. Currently, there are more controlled studies to test people during their NDEs.

For this book we will define heaven as the survival of consciousness after the death of the body. We might not be able to build a machine to detect heaven, but we may already possess a detector—the human brain. With time and new technology, we may be able to document the survival of consciousness.

SECTION II
BELIEF IN HEAVEN

Chapter 4
Ancient Spiritual Beliefs

To test the heaven hypothesis, it's important to examine human beliefs about heaven, spirits, gods and the transition between life and death. We don't know exactly when the belief in spirits and an afterworld began, but it seems that every culture has its own views. We will start this section by looking at the oldest beliefs developed by cultures in ancient times.

Animism

Animism is the oldest known type of belief system and predates paganism. It asserts that all things, including animals, plants, rocks, rivers, and weather are spirit filled. It is still practiced by many traditional societies in a variety of forms. Their beliefs are so widely accepted that they often do not even have a word in their languages that corresponds to animism or religion.

Many religions include some element of animism combined with additional beliefs. The present-day Berbers of northwestern Africa practice several different religions. Their traditional, ancient religion included shamanism, polytheism, and animism. Many Asian religions include animism as a dominate theme. Buddhism has a significant emphasis on coexisting with nature. Sikhism originated in India, and like Buddhism, believes that humans are spiritually sensitive to the natural world. Other Indian religions focus on worshipping specific trees, and revering sacred rivers, mountains, and landscapes as part of their beliefs system. In the Philippines, the Anito not only believe that natural objects such as grass, trees, and rocks have their own spirits, they believe there is an alternative spiritual world that exists alongside the physical one.

Believing that all organisms might have spirits seems a bit strange to those of us in the western world. How can a tree be a sacred being? It's just a plant, isn't it? We think of plants as living organisms that sit around drinking up sunlight, but recent research has shown that they are much more dynamic. Maybe those who believe in animism have been one step ahead of the scientific community. Plants have been shown to communicate with each other and a host of other organisms using organic compounds, electrical signals, and through mycorrhizal (fungal) networks growing through the soil. Plants also respond to sound, and some produce ultrasonic sounds that can be heard by bats, mice and some insects. Maybe the belief of plant having spirits isn't as far-fetched as it seems.

Ancient Africa

Since Africa was the birthplace of humankind, one would expect the oldest beliefs to be represented there. Of course, the continent is as diverse as its 50 countries, with hundreds of ethnic groups and over 1000 languages. New religions were also brought in with invading Christians and Muslims, who attempted to replace the old ways of thinking. Despite that, there are some general beliefs which still survive today.

According to the old religions, the spirit world is integrated into the living realm. All living and non-living objects are sacred at some level, which affects all aspects of life, including economics and politics. The sacred and the secular influence each other and there is no question to believers that spirits influence the workings of the world. Codes of behavior are established by spirits and troubles can be caused by the actions of negative spirits. For example, the Bantu in southern Africa believe that illness is caused by imbalances in the relationship with the spirits.

Their myths, stories and beliefs are flexible, changing with the times. That may be because traditions were passed on orally, not written and seen as unchanging. While the universe had a beginning, traditional belief says that it does not have an end. The earth is also considered a living thing, a goddess. The human world is located between the sky, which holds gods and higher spirits, and the underworld. The universe is an active, ever-changing system. Agents from all the realms are interacting with each other.

Many beliefs include a supreme god which may be a remote or absent creator god, but most of the time this god is equal to other spirits, not above them. Power originates from God, who is believed to dwell in the skies. This power is passed on to a hierarchy of divinities, spirits, and the living dead. It is also available to some people in varying degrees. Since the border between worlds is porous, diviners and priests are used to relay information through dreams and divination.

Spirits exist in a pantheon, usually of 201 or 401. The spirits that a person reveres are often inherited from their family and are a subset of all the spirits. Ancestors are also important. They are equal to deities. Some are so old they were not known by those living. They are said to transcend to the higher realm where they can bestow honors and blessings, and they are used to assist in all rites of passage.

Although they believe in the continuation of life after death, the spirit world of the ancestors is in some ways like life on the earth where there is a pursuit of plenty. There is no ultimate reward and punishment promised, but the ancestors do need people to sustain them in the afterlife.

Australian Aboriginals

The Aboriginal People of Australia are a special group who bridge the time between ancient and modern spiritual belief. DNA suggests that their ancestors left Africa about seventy thousand years ago. Since Australia was isolated from much of the world, their culture is the oldest continuous one on the planet. Ancient Rock Art can be seen as engraved geometric forms, such as circles, concentric circles, arcs, dots or animal tracks. There are also painted or engraved silhouettes of humans or animals. Petroglyphs were created on almost any stone surface, while pictographs usually survive only in sheltered areas. Rock art locations are considered sacred sites and are found across the continent. Today, modern artists carry on the tradition, painting some of the same designs, only on bark, paper and wood.

An important concept for the Aboriginal people is Dreamtime. I'm not sure if people who are raised outside of their community understand it entirely, but it is the heart of their culture and intrinsically linked to the land and their cave art. Dreamtime is often referred to as the beginning of knowledge from which laws of responsibility and survival sprang. There are sacred locations associated with Dreamtime where seasonal rituals were performed, and stories told. Although they differ from region to region, many include tales of mythical ancestral spirits who created the land and the sky.

Cave art symbols are not only related to the tales told during ceremonies they are also important clan representations. Clan symbols are made of a set of lines drawn using specific ochre colors that represent elements such as fire and water and a totemic animal design. The two features indicate which clan created the artwork and the person's identity. The symbols can also contain deeper meanings and interpretations.

Belief in the afterlife is similar across the continent. There is no heaven or hell and a person's actions during life have no influence on his spirit in the afterlife. When a person dies, they travel to the sky or other locations where they stay with the creative beings and the beings from the Dreamtime. Death is part of a cycle. The person retains their individuality just after death, but that is a temporary state. As time passes, the person's identity fades and becomes eternal. Part of the spirit returns to its home site until it can be reborn.

Native North American Nations

As archeologists find more ancient sites in North America, it looks like humans may have arrived on the continent twenty thousand years ago, by crossing the Bering Land Bridge or by sailing from Asia to the western coast of North America. So, we can assume that the American people, like the Australians, were separated from outside influences for at least that long, and their beliefs in the afterlife originated from ancient traditions.

When Europeans invaded North America, there were an estimated 500 Indigenous religions and/or spiritual beliefs. It is difficult to make broad generalizations about their traditional viewpoint about death. Europeans who recorded indigenous beliefs did so through a Christian frame of reference. Thus, we ended up with terms like the "happy hunting ground." It's interesting to note that in the several hundred distinct Indigenous languages, there was no world which translates as "religion." Like other ancient beliefs, the spirit world and death were part of life.

Beliefs in the afterlife vary greatly from tribe to tribe. For many, those who have passed to the spirit world can occasionally communicate with the living through dreams or medicine people. Others believe in a more structured land presided over by a supernatural caretaker. The dead may become stars or part of the earth unless they are disturbed and turned into ghosts. For other tribes there is no afterworld. The spirits of the dead are reincarnated as newborns within their family or clan.

Death, and what happens afterward, is not a focus. They are practical in that they feel they have no way of knowing, so they do not waste time thinking about it. In their classic 1911 ethnography, *The Omaha Tribe*, Alice Fletcher and Francis LaFlesche report: "There does not seem to have been any conception among the Omaha of supernatural rewards or punishments after death."

Burial customs varied across the tribes. They included cremation, where burning helps the deceased reach the afterlife. Some placed the ashes in a mortuary pole as a memorial. Western tribes practiced tree burial, letting animals consume the body, like a sky burial. And last, there was an earth burial where the body is returned to mother earth.

Symbols etched onto the sides of boulders lying along the western edge of a desiccated lake in Nevada are the oldest confirmed rock carvings in North America. Researchers concluded that the petroglyphs, located about 35 miles northeast of Reno at Winnemucca Lake, are at least ten thousand years old. The petroglyphs range from simple lines, pits, and swirls to more complex and ambiguous shapes that resemble diamonds, trees, flowers, and veins in a leaf. Their meanings are unknown.

Natural drugs were also used to create spiritual enlightenment. The modern Peyote Religion stems from a tradition that is thousands of years old, thought to have originated within the Apache or Cree tribes. It uses the hallucinogenic, *Lophophora williamsii,* cactus as part of ceremonial and sacred rites.

Chapter 5
First Words about the Afterlife

Sumerians

The earliest writing about the afterworld was found in Mesopotamia. Cuneiform, angular shaped text, was developed by the Sumerians about fifty-five hundred years ago. Along with writing, they developed wheeled vehicles, codes of law, a calendar, mathematics, astronomy and medicine. They built the first city-states, including Uruk, Ur and Eridu along the Tigris and Euphrates rivers, which are now located within the borders of Iraq.

There are no specific records that explain Sumerian beliefs in gods and the afterworld, but a variety of written accounts can be pieced together to create a whole. The *Epic of Creation* describes the god, Marduk, killing Tiamat, the mother goddess, the earth's creation, paradise of the gods, and the netherworld. The Sumerians believed that an afterlife existed in a land below our own. Everyone went there after death. There was no heaven or hell depending upon one's actions on earth. The netherworld was a world of weak and powerless ghosts who dwelled in darkness. Other stories said that due to an error, all humans must die and only the gods attained everlasting life.

Egyptians

The Egyptians began using hieroglyphs over five-thousand years ago, about the same time that the Sumerians developed Cuneiform. They believed that their writing was invented by the gods, who were a major influence in their everyday lives, and that writing helped spread their religious beliefs throughout Egyptian society. That included the creation of the *Egyptian Book of the Dead* which illustrated and explained what the dead would need to know in order to complete a safe journey to the afterlife.

Egyptians believed in an underworld, eternal life, and rebirth of the soul. To reach the afterlife, many practices had to be performed during one's life, including following the beliefs of the Egyptian creed. The creed emphasized harmony, following the principles of the goddess of truth, Ma'at, and living justly. They were required to maintain high religious morals and comply with a variety of traditions.

The underworld, known as *Duat,* had one entrance reached by traveling through the deceased's tomb. The path taken to the afterworld may have differed between kings and common people, but for both, rituals were completed after the individual's life ended. The soul would enter corridor lined with an array of statues, including a representation of Horus, the hawk-headed god. They would next meet Osiris who would determine the virtue of the deceased's soul. If the soul passed judgement, it was granted a peaceful afterlife. Eternal life wasn't seen as staying in the afterworld but instead being reborn indefinitely. Souls of those who had lived a good life were guided by Osiris to be born again.

In the Chapter of Coming Forth and Passing Through in *The Book of the Dead* it is written:

The plug has been lifted from the unguent jar. A perfume of hours. The past has been rolled into a scroll I shall not see again. The eye of the hawk is unblinking. Open. Shut. Perfect.

I rise like the sun above olive trees, like the moon above date palms. Where there is light, I shall be. Where there is darkness, there is none of me. I rise like the moon above date palms. I am counted as one among stars.

Beam of light, sun and moon. Shining beast, man and woman. I am passing through. Come outside among the people. I am light. Gaze on me. Moon in darkness, sun in morning. Light is what I will on earth, along the Nile, among the people.

I have traveled through the tomb, dark and lonely ground. I am here now. I have come. I see. In the underworld, I embraced my father. I have burned away his darkness. I am his beloved. I have killed the snake. I have given him meat. I walk in my sleep through earth and heaven.

I have set the sky in two parts. I pass through. I wander the horizons. I have dusted my feet with earth. I have worn the skin of a black panther and chanted into the ears of children. I eat with my mouth. I chew with my jaw. I am a living god come forth. I am with the earth millions of years.

The Egyptians had a very complex view of the afterworld and how a person could reach it. They were not alone in their complexity of thought.

Hinduism

Hinduism claims to be the oldest surviving religion on earth and is the world's third largest. Most scholars believe it began sometime between forty-three and thirty-five hundred years ago in the Indus Valley, near modern-day Pakistan. Hinduism had no founder but is a fusion of various beliefs. This may be why there is a diversity of opinion and authority. It is a collection of many faiths under one umbrella, and a person is free to choose a preceptor to study under. They follow the tenet: *Truth is one, paths are many.* Currently, the four major denominations of Hinduism are Smarta, Vaishnavism, Shaivism and Shaktism.

The *Shruti* is the basis of most ancient religious texts, believed to be eternal knowledge transmitted by sages. These include the four Vedas with sections: the Samhitas, Brahmanas, Aranyakas and Upanishads.

Hinduism has four basic principles that are shared by most in the faith. First is Karma, a law of cause and effect that states that a person's actions come back to them. Second is Reincarnation, a natural cycle that brings the soul back to earth after death. Third is Dharma, the divine law of the universe that maintains order. Fourth is an all-pervasive divinity or God that is timeless and formless.

The Hindu belief in the afterlife centers around their belief in reincarnation. The afterlife is usually a short interval between when a soul dies in the current incarnation and when it is born into a new body. Humans are in a cycle of death and rebirth called *Samsara.* Some believe that the soul may enter *swarg* (heavenly realm) or *narak* (hellish realm) for a period before rebirth. Ultimately, the soul is trying to reach *moksha*, the level of attainment in which the soul is freed from the process of reincarnation. Hinduism strongly encourages cremation. It not only helps with the disposal of the body but also aids in ushering the soul into the afterworld and reaching moksha.

Abrahamic Religions

The Abrahamic religions are those who trace their linage back to the patriarch Abraham of the *Bible*. Its first five books: Genesis, Exodus, Leviticus, Numbers, and Deuteronomy reached their present form about 2,500 years ago. Their authors were believed to be the elite who controlled religious thought at that time. The books describe the history of the Hebrews and contain other stories which may derive from older Sumerian sources. For example, the Genesis flood story is very much like the Gilgamesh flood myth. The original story of Moses is like the story of Sargon of Akkad.

In the books, the Hebrew word *Shamayim* is translated as heaven. It is defined as a region above the earth, the heavens, the place of the stars, sky, and air, or the invisible realm of God. There is no description of how the Hebrews viewed heaven in the Scripture. In the books of *Psalms* and *Isaiah*, God's throne is said to be in heaven. In *Kings*, Solomon says that heaven is God's dwelling place. Along with heaven, there also existed the realm of the dead, *Sheol*. The word means "the place of the dead" or "the place of departed souls/spirits." This is where the dead went to rest, not *Shamayim*.

Judaism

Jews call the collection of sacred Hebrew works from Moses to Ezra the *Tanakh. Tanakh* does not comprise the totality of Judaism, but it does provide the roots and origins of everything. Reading through *Tanakh,* there isn't much about heaven or an afterlife. It appears that they were not focused on getting to heaven. Instead, they were concerned about bringing heaven down to earth.

Heaven was not a supernal realm one goes to after death. Heaven was what happens on earth as the fruit of our collective labor. Early rabbinic texts describe *Olam Ha Ba* as an idyllic version of this world. It is a physical realm that will exist at the end-of-days after the Messiah has come, and God has judged both the living and the dead. Bad people would be sent to *Gehenna,* a place of punishment, but only for 12 months. Then they would go to *Gan Eden,* a paradise on earth for those who had lived a righteous life. They believed that those who slept in the dust of the earth (*Sheol*) would awaken, some to eternal life, others to reproach or everlasting detestation.

Early religious leaders not only taught that heaven was a place only for God, but it was also off limits for mortal people. There might be a reason for this. In Hebrew burial caves dating to about four thousand years ago, jars and bottles of liquids and oils, small plates of food, weapons and jewelry were found with the dead. Ancient Hebrews were more like their neighbors in Canaan. They believed that dead relatives lived in S*heol* after death, and they needed to be taken care of. To discourage ancestor worship, Jewish leaders eliminated the underworld.

Christianity

Christianity revolves around the life, death and resurrection of Jesus Christ, the son of a Jewish carpenter, who questioned the teaching of his faith and preached a new message to the masses. Christians believe that God sent his son Jesus, the Messiah, to save the world. Jesus was crucified on a cross to forgive of sins of mankind and was resurrected three days after his death before ascending to heaven.

It was Christians who brought the afterlife back to the children of Abraham. Jesus came proclaiming the coming of a spiritual kingdom that he would rule over. In the *Bible*, John 14:2, it says that in Jesus' Father's house there are many mansions. In Corinthians, Saint Paul shares his experience of seeing the third heaven, the dwelling of God. Paul said he received the vision of heaven fourteen years before writing the second letter to the Corinthians. That would mean the vision occurred around 43 AD.

Paul described himself as a man "in Christ who fourteen years ago was caught up to the third heaven" (2 Cor 12:2). He was "snatched up," but said that he did not know if he was in or out of body during his revelation (2 Cor 12:2, 3). He claimed to have "heard unsayable sayings," as he tried to describe the experience. Paul gives little information in 2 Corinthians, but we know that Paul somehow was caught up to the third heaven or paradise. According to the account, paradise is where Jesus is, Paul heard unsayable sayings, and those sayings may have come to him as a heavenly or angelic dialect.

That was all that was included in the Biblical account, but in the shorter text, *The Apocalypse of Paul*, discovered in the mid-1940s as part of the library at Nag Hammadi, God took the form of a little child and spoke to Paul. They traversed not only third heaven but ascended all the way to the tenth heaven. In third heaven, Paul was still able to see the earth below. Heavens four and five were the realm of angels. Number six was the place of a "great light" and number seven was the place where Paul saw "an old man." When Paul reached level ten, he was united with "fellow spirits."

In the New Testament, *Sheol* became *Hades (Hell)*, which is also a general reference to "the place of the dead." *Hades* is divided into a place of blessings and a place of torment. The souls of the righteous, at death, go directly into the presence of God—the part of *Sheol* called "heaven," "paradise," or "Abraham's bosom" (Luke 23:43, Corinthians 5:8, Philippians 1:23). It also seems to be a temporary place where souls are kept as they await the final resurrection.

Unlike the Jews, Christians divided the afterlife into two regions, heaven for the good, and hell for the bad. In some verses in the Book of Job, *Hades* is described as a land of the deepest night. It is also described as a place with everlasting fire. Mark 9:43 also gave the description of "unquenchable fire." Jesus warned in many instances about the suffering in *Hades.* The expression "gnashing of teeth" was used to describe the extreme pain that would be endured.

Islam

Islam is the most recent of the Abrahamic religions. It began with the Prophet Muhammad who he received his first revelation in 610 AD. Muslims believe the archangel Gabriel revealed the word of God to him in Arabic. The revelations were collected and codified as the *Qur'an.*

In Islam, paradise or the garden (*Jannah*) is the final land of the righteous. Belief in the afterlife is one of the six articles of faith in Sunni. While believers go to *Jannah,* non-believers go to *Jahannam.* Both locations have several levels, *Jannah* being desirable and

Jahannam being a place of punishment. Although humans may not pass the boundaries to the afterworld, the afterworld may interact with the temporal world of humans. Many scholars believe that there is the eighth level of heaven. A few even believe that there are a hundred levels of heaven in Islam. Hell awaits those who reject Allah or cause mischief on the earth. It is described as a miserable existence of constant suffering and shame.

Heaven is a reward for good works.

> *And give good tidings to those who believe and do righteous deeds that they will have gardens [in Paradise] beneath which rivers flow. Whenever they are provided with a provision of fruit therefrom, they will say, 'This is what we were provided with before.' And it is given to them in likeness. And they will have therein purified spouses, and they will abide therein eternally. (Q. 2:25)*

Heaven is described as a garden.

> *Those will have gardens of perpetual residence; beneath them rivers will flow. They will be adorned therein with bracelets of gold and will wear green garments of fine silk and brocade, reclining therein on adorned couches. Excellent is the reward, and good is the resting place. (Q. 18:31)*
>
> *...the chosen servants of Allah. Those will have a provision determined—fruits; and they will be honored in gardens of pleasure on thrones facing one another. There will be circulated among them a cup from a flowing spring, white and delicious to the drinkers; no*

bad effect is there in it, nor from it will they be intoxicated. (Q. 37:40-47)

[Other] faces, that Day, will show pleasure. With their effort [they are] satisfied. In an elevated garden, wherein they will hear no unsuitable speech. Within it is a flowing spring. Within it are couches raised high and cups put in place and cushions lined up and carpets spread around. (Q. 88:8-16)

Other Religions

Many other religions came into being at similar times or after these older beliefs were established. The Greeks and Romans worshiped a pantheon of Gods who were able to intermingle with people in the earthly realm. The Gods lived in the upper heaven, a place of lightning, omens and the sun and planets. A pantheon of gods existed across northern Europe. The Norse, for example, believed in Valhalla, a heavenly hall presided over by the god Odin. There were five realms that the soul could travel to after death.

In Asia, Confucianism, Taoism and Buddhism followed Hinduism. Buddhism sees heaven as a temporary realm in a cycle of rebirth to achieve enlightenment. Taoism describes multiple heavens inhabited by gods and immortals which is important to both personal and social development. The beliefs of the Mayans, Incas and Aztecs in Central and South America took hold later. They believed in multiple gods and multiple levels of the afterworld. People who lived a good life would ascend there after death.

Most religions include some type of belief in an afterlife. Since this book is not a review of religions of the world, I will not take any more time to discuss these other interesting and important belief systems.

Chapter 6
Reincarnation Beliefs

Reincarnation is a belief that spans several religions. I will include it here as a separate topic because if a spirit/soul returns to Earth for another lifetime, it could present a situation which can be scientifically evaluated. If a person retains the memory of a past life, careful documentation may demonstrate that they actually lived another life before their present day one.

As mentioned above, Hinduism was one of the earliest religions that hold reincarnation as a core belief. Other Asian religions that believe in reincarnation include Buddhism, Jainism and Sikhism. Reincarnation is also part of Abrahamic beliefs including Kabbalistic Judaism and Druze.

Buddhism

Reincarnation is a central concept in Buddhism. It is known as *Samsara*, the cycle of life, death, and rebirth. In Buddhism, this cycle is not a simple journey from one life to the next. Instead, it is part of a deeper process that includes all living beings. It involves the mind, karma (the effect of a person's actions), and the nature of suffering. Buddhists believe that all living beings are trapped in the continuous cycle of *Samsara*, unless they take action to escape it. The journey is considered to be eons long for each soul, with the traditional number of rebirths to be 8,400,000. The goal in Buddhism is to exit the cycle and attain *Nirvana,* a state of liberation and peace.

Many in the west believe that Buddhist reincarnation is the migration of a soul into another body after death. There is no such teaching in Buddhism. An important doctrine of Buddhism is *Anatta* which means "no soul" or "no self." Unlike with Hinduism, there is no permanent essence of an individual self that survives death. What survives is the continuation of a stream of consciousness, which is shaped by karma from previous lives. *Nirvana* is achieved when followers realize that there is neither self nor consciousness. It is described as the extinguishing of the 'fires' of attachment, aversion and ignorance that cause suffering.

The Dalai Lama wrote:

> *Therefore, Buddhists believe that there is an end to being reborn as a result of karma and destructive emotions, but most Buddhist philosophical schools do not accept that the mind-stream comes to an end. To reject past and future rebirth would contradict the Buddhist concept of the ground, path and result, which must be explained on the basis of the disciplined or undisciplined mind. If we accept this argument, logically, we would also have to accept that the world and its inhabitants come about without causes and conditions. Therefore, as long as you are a Buddhist, it is necessary to accept past and future rebirth.*
>
> *For those who remember their past lives, rebirth is a clear experience. However, most ordinary beings forget their past lives as they go through the process of*

> *death, intermediate state and rebirth. As past and future rebirths are slightly obscure to them, we need to use evidence-based logic to prove past and future rebirths to them.*

Jainism

Jainism is an Indian religion that teaches spiritual purity can be achieved through non-violence expressed to all living creatures. Their three ethical pillars are: *ahimsa* (non-violence), *anekantavada* (truth and reality are many-sided) and *apangraha* (no-possessions). The goal is to achieve *moksha* or liberation from karma.

Like Hindus and Buddhists, Jainists follow *Samsara,* the cycle of birth, death and rebirth. Souls can be reborn into one of four states: heavenly beings, humans, animals/plants and hellish beings. The universe is divided into three realms: the *Urdhva Loka* (upper world of heavenly beings), the *Madhya Loka* (middle world of humans, animals, and plants), and the *Adho Loka* (lower world of hellish beings). Only a soul with a human body can attain enlightenment and liberation.

Lord Mahāvīra (born 599 BC) said:

> *Attachment and aversion are the root cause of karma, and karma originates from infatuation. Karma is the root cause of birth and death, and these are said to be the source of misery. None can escape the effect of their own past karma.*

Sikhism

Sikhism was founded in the Punjab region of south Asia in the 15th century. Their core beliefs include a belief in one God, that all humans are equal, that all should engage in selfless service, and the three pillars (remember God through prayer, earn an honest living, and share one's earnings with others). The religion also teaches that the soul reincarnates when the body dies. If one does not perform righteous acts, one's soul will continue to cycle in reincarnation for eternity. A person who has performed good deeds in their life is transmigrated to a higher life form in each succeeding life and eventually becomes Godlike.

Sikhs believe that good or bad actions in this life determine the lifeform into which a soul takes rebirth. At the time of death, ego-centric souls may be destined to suffer great agonies and pain in the dark underworld of *Narak.* A soul who achieves grace overcomes ego by meditating on God and is liberated from the cycle of reincarnation. The emancipated soul experiences salvation in *Sachkhand,* the realm of truth.

The *Guru Granth Sahib* is Sikhism's holy scripture. It is a collection of hymns, poems, ballads and writings covering 1430 pages and written by 43 authors. Author Bhagat Trilochan wrote on the afterlife. He said that at the time of death the final thought determines how one reincarnates.

> *The soul takes birth in accordance with what the mind last remembers. Those who dwell on thoughts of riches or worries about wealth are born again as serpents and snakes. Those who dwell on thoughts of carnal relations are born into brothels. Those who remember their sons and daughters are born as a pig to become a sow birthing a dozen or more piglets with each pregnancy. Those who dwell on thoughts of their houses or mansions, take the form of a ghostly goblin*

type specter haunting houses remembered. Those whose final thoughts are of the divine, merge eternally with the Lord of the Universe to dwell forever in the abode of radiant light.

Translated by Sukhmandir Khalsa, Learn Religions, April 9, 2019, www.learnreligions.com

Kabalistic Judaism

Among the most prominent forms of Jewish mysticism is Kabbalah, which emerged in the 12th century. Followers believe that the Torah is a multi-layered document. Many of its deeper interpretations are not obvious to the untrained reader and do not deal with the practical application in daily life. As such, they are not of interest to most of the Jewish population. One of those deeper meanings is *gilgulim* (reincarnation), the "revolving" of souls through a succession of lives. It is an integral part of Jewish teaching that has always been present but isn't followed today.

Ramban, one of the great commentators on the Torah and a seminal figure in Jewish history, hints several times that reincarnation is the key to penetrating the deep mysteries of the *mitzvah of yibum* (the obligation of the brother of a childless, deceased man to marry the widow). In his explanation of Genesis 38:8, he states that Yehudah and his sons were aware of the secret of reincarnation, and that this was a major factor in their attitudes towards Tamar.

Reincarnation does not pre-determine one's life. It is, like with other faiths, an opportunity for restoration and soul-perfection. The holy Ari explained that every Jew must fulfill all 613 *mitzvot. Mitzvot* are commands such as having children, declaring God's oneness, not eating pork, obeying the sages and providing interest-free loans. If a person fails in one lifetime, he comes back again and again until he finishes. For this reason, events in a person's life may lead him towards certain places or encounters that do not make sense. These are opportunities the person needs to fulfill those particular *mitzvot* necessary for the perfection of his soul.

Many scholars argue that these mystical concepts can easily be misunderstood or carried to misleading conclusions. That is probably why reincarnation is only hinted at in scripture, and why some knowledge and a great deal of determination are often required to gain access to this information.

Druze

Followers of the Druze religion are mainly found in Lebanon, Jordan, Syria and Israel, Although the Druze faith developed from Islam, Druze are not Muslims. The Epistles of Wisdom, influenced by several religions, is their central text and emphasizes the mind and truthfulness. They maintain their Arabic language and culture, but most Druze keep their religious practices secret. Conversion to their religion is not allowed and interfaith marriages are rare.

The Druze believe that God is the "whole of existence." They differentiate between spiritual individuals, known as *uqqāl,* who keep their faith's secrets, and secular people, known as *juhhāl,* who focus on worldly matters. A human soul will transfer only to a human body, in contrast to other belief systems. Additionally, a male Druze can be reincarnated only as another male Druze and a female Druze only as another female Druze. A Druze cannot be reincarnated in the body of a non-Druze. Souls cannot be divided, and the number of souls existing in the universe is finite. Druze believe that, after completing the cycle of rebirth through successive reincarnations, the soul reunites with *al-'aql al-kullī,* the cosmic mind.

Druze believe in a practice called, *notq,* remembering and talking about one's past life. Since it is important for the Druze religious community, there have been studies examining its psychological impact. Some researchers focused on 'solved' stories, ones in which the person identifies his/her previous incarnation. Others have dealt with stories that are "open" or unsolved. Researchers were not testing the validity of the memories but were studying the psychological impacts of past life memories. They have found that despite its ambiguity and many unsolved questions, *notq* does not harm the individual psychologically and may contain therapeutic value due to being consistent with the Druze beliefs.

Chapter 7
Spiritualism, Mediums and the Afterlife

Spiritualism

Although the belief in talking to ancestors and the existence of the spirit world had been ongoing for centuries, modern Spiritualism did not begin until the mid-1800s. In the United States, the story of Spiritualism began with a small cabin in a rural town in western New York State. The Weekman family lived in the cabin during 1846–47. At the time, they heard footsteps in the cellar and loud raps on the floors, doors and walls. They attempted to catch the perpetrator but were never successful. At one point, their eight-year-old daughter was awoken in the night by feeling something cold touching her face. Mr. Weekman heard his name called when he was there alone. Their servant saw the apparition of a man in the bedroom. Needless to say, the Weekmans did not stay in the house long.

In December of 1847, John David Fox with his wife and two daughters, Catherine, aged twelve, and Margaretta, aged fifteen, became the next tenants. The family was frequently disturbed at night with the sounds of furniture being moved as well as knocking on the doors and walls. Later, they heard gurgling, as if someone was being choked, and the thump of a body falling to the floor. At one point, the girls were frightened by a large weight lying across their feet and a cold hand passing over their faces. It became impossible for them to sleep.

Instead of running, the family challenged the entity. The girls soon discovered that they could communicate with it. They taught it to knock in response to their questions in a code type format. Using the knocking responses, they discovered that the spirit was a man who had been killed and supposedly buried in the cellar. The entire story was recorded at the time in *A Report of the Mysterious Noises Heard in the House of Mr. John D. Fox in Hydesville, Arcadia, Wayne County.*

The girls eventually moved to Rochester, New York, where they used their knocking technique to speak to other spirits. The interest in talking to spirits soon became popular and spread across the United States and Great Britain. One of its early proponents was Emma Hardinge Britten. Britten was born in England in 1823 and was an influential leader in establishing early Spiritualism. She was a passionate orator and writer on the spiritual movement, as well as a practitioner. She published two significant books, *Modern American Spiritualism* (1870) and *Nineteenth Century Miracles* (1884), which are detailed accounts of the early history of the modern spiritualism movement in America.

According to Britten, Spiritualism is "the discovery of a mode whereby discarnate spirits can and do communicate systematically with earth." It teaches that people are, first and foremost, spirits. The body is a material vessel in which the spirit grows and becomes an individual. Once the body dies, the spirit continues.

Communication with the spirit world or heaven is the foundation of Spiritualism. Spiritualists do not rely on faith alone, like many other religions. The medium's goal is to demonstrate the existence of a spirit world by contacting specific individuals who have passed from the earthly realm. The medium identifies the person who has passed and offers proof to their living family and friends of their existence.

Over the decades, untold spirits were reached by mediums. Their names, ages, dates of birth, important events and secrets were revealed to loved ones. They communicated by knocking or rapping, appearing in photographs, creating paintings, and writing out their desires. Spirits also directed and inspired healers. Britten said, "they have presented wise and wonderful visions, and in thousands of ways blessed, benefited, and assisted the friends they have left behind."

According to Britten, Spiritualism is a means of communication, but it does not originate as an earth-based theory or opinion. Mediums relate what they sense; none of their authority comes from books written by men. Spiritualism is not a religion per se, in that it does not need endorsement from any fixed creed, sect or society. It, instead, requires belief through demonstration, as opposed to blind faith. It may be in harmony with some established religions and even the laws of science, but functions independently from them. Spiritualism proceeds "by virtue of spiritual laws, absolute and true, whether man understands them or not."

Several spiritualist mediums received more than messages from loved ones. They were given glimpses of heaven or the spirit world. Two examples include G. Vale Owen and William T. Stead.

Reverend G. Vale Owen

G. Vale Owen was born in Birmingham, England in 1869. He was educated at Midland Institute and, in 1893, ordained by the Bishop of Liverpool. He worked as a curate in several parishes in Liverpool until 1900 when he became vicar of Orford.

Owen's psychic abilities manifested after his mother's death in 1909. He began to receive communications via automatic writing from her in 1913. Automatic writing is written by a medium while conscious or in a trace state with pencil/pen and paper. Messages are believed to come not from the mind of the medium, but from a spirit controlling the hand of the medium to deliver a message. In Owen's case, he was the medium. The writing impacted him so significantly that he converted to Spiritualism, much to the dismay of his bishop. Church authorities were unhappy with Owen's spiritual practices and forced him out of his parish in 1922. At that point, he began to actively promote Spiritualism, and toured England and the United States.

The messages Owen received from his mother were transcribed into articles for *The Weekly Dispatch* and then into a series of books during the 1920s. *The Life Beyond the Veil* was composed of four books: *The Lowlands of Heaven, The Highlands of Heaven, The Ministry of Heaven, and The Battalions of Heaven. Outlands of Heaven* contained two books: *The Children of Heaven* and *The Outlands of Heaven.*

The first entry in *The Lowlands of Heaven* is dated Tuesday, September 23, 1913. His mother's spirit said that she lived in the first sphere of heaven, where recently passed souls arrived. She described it as "earth made perfect." The level does not have a fourth dimension which made it difficult for her to describe adequately, but there are "...hills and rivers and beautiful forests, and houses, too...." Her major concern was to prepare the place for the arrival of others.

She went on to describe a series of crystal halls of different colors. They were places to study the effects of color on the physical world. "Our environment is part of us," she said. "Just as with you, and light is one component, and an important one, of our environment. Therefore, it is very powerful in its application, under certain conditions, as we saw it in these halls."

Owen's description of the spirit world continued with communications from Zabdiel, a spirit who resided in the tenth level of heaven. Zabdiel made it clear that the numbering system for the levels of heaven is an artificial construction, but it is easily understood by people living in the physical world. Zabdiel's main occupation was helping others ascend.

Zabdiel said, "For not until a man has learned to love all without hating any is he able to progress in this land where love means light, and those who do not love move in dim places where they love their way, and often become so dull in mind and heart that their perception of the truth is as vague as that of outward things."

A new group of spirits came forward to present information to Owen and were published in *The Ministry of Heaven.* Their stories were similar to Zabdiel's. There was also an account of a mission down to the Lower Zone of the spirit world, where they assisted those who were experiencing intense suffering.

Owen's writings impressed Sir Arthur Conan Doyle, a famous Spiritualist at the time. When questioned about Owen's legitimacy, Conan Doyle wrote a letter to the Editor of *Light.* He told people who thought Owen was just trying to make money that, "It is not sufficiently known that Mr. Owen refused to take any remuneration for the Script when it was published in Lord Northcliffe's paper. Also, that he worked without any profit for himself in America. Since then, he has had one long lecture-tour in the provinces which did much good, but which placed a great strain upon his health without any corresponding pecuniary return."

Owen eventually became pastor of a Spiritualist congregation in London. By that time, his financial resources were severely depleted. Conan Doyle organized a collection for him and created a trust fund that provided him with financial support for the rest of his life. In 1931, Owen he fell ill, probably exacerbated by the strain of mediumship. He died in March of that year at the age of 61.

William T. Stead: *The Big Blue Island*

When the Titanic sank in 1912, Estelle W. Stead and Pardoe Woodman were on tour with a Shakespearean Company in Europe. On the same day that the Titanic sank, Pardoe had a vision of a sinking ship. He described the sinking to Estelle while they were having tea. He did not know the name of the ship nor a man he called William, but he described them both. It wasn't until later that Estelle was informed that her father, William T. Stead, had died during the sinking.

A few weeks later, Estelle, while sitting with well-known direct voice medium Etta Wriedt, saw her father's face and heard his voice telling her goodbye. She talked to her father's spirit for over 20 minutes. The session was recorded and published in an article in *Nash's Magazine.*

In 1917, Pardoe was discharged from the army as an invalid and came to stay at Estelle's country cottage at Cobham. He had lost a good friend in the war and was interested in communicating with him. When local mediums arrived to help him communicate with the friend, he found that William was also there. William was helping facilitate communication from the other side.

Soon after, Pardoe found that he had a natural ability for automatic writing (subconscious writing completed under the direction of a spirit). William and other spirits were able to communicate through him by this method. During 1918, Pardoe conducted writing sessions regularly each week. William first kept them informed about what was happening on the war front. After the war, William wanted to send them messages about the other side. They sat once a week for several weeks. Pardoe wrote down William's messages and published them as *The Blue Island.*

In the *Big Blue Island,* William described his first surprises. First, he was shocked that he was in a position to help people. Second, he found several friends that had passed before him. He saw people as he knew them in life. He described the heaven that he experienced.

> *Another thing which struck me was the general coloring of the place; of England it would be difficult to say what the impression of coloring is, but I suppose it would be considered grey-green. Here there was no uncertainty about the impression; it was undoubtedly a blue which predominated. A light shade of blue. I do not mean the people, trees, houses, etc. were all BLUE; but the general impression was that of a blue land.*

He explained how his small group of people were comprised of all colors, races and sizes who lived freely together. Most he did not know. People were tending to their own needs, and all were at peace. There was a sea nearby and a beach to walk on. He spoke of eating meals and seeing buildings. This was apparently necessary for newly arriving spirits so that they would feel comfortable in their surroundings. "It seems that all the senses are provided for here. The chief work on this island is to get rid of unhappiness at parting from earth ties, and therefore, for the time being, the individual is allowed to indulge in most of earth's pleasures."

He also said that the island was a place to shed the physical world.

> *The process of discarding is a gradual one. As we live here we gain knowledge of many kinds, and come*

> *to find so many things, hitherto thought essential, not only of no importance but something of a bore, a nuisance, and that is how we grow to a state of dropping all earth habits.*

William described in several chapters his attempts to contact the living and how others contact those they have left behind. He said he learned there were other locations in the afterworld.

> *These other lands are nearly all inhabited by a higher form of life, a happier form and, individually, a more powerful form, but there are one or two other lands of not so high an order, where happiness is less or not at all, according to whether life on earth was a well or lightly ordered thing. In these lands the people who are there fail and fail again to find the spirit in themselves to desire to rise, to improve and control themselves, although the necessary strength is offered and offered and even thrust at them.*

He said the blue island is a temporary location to acclimatize the newcomer. Spirits can move on to the "real world." In this world the color is "toneless," and people are engrossed in regular routines. There are still buildings and people can choose where to live. "When we come to this land, we have ceased to desire food, drink and sleep; we are now pure spirit in the rough state; there is still more refining to be done in the next phase." He emphasizes the degree of freedom that they have. "Life here is a grander thing–a bolder thing, and a happier thing for all those who have led reasonable lives on earth. But for the unreasonable there are many troubles and difficulties and sorrows to be encountered."

Spiritualist Investigations

Unlike most religions that rely on faith to support their belief in heaven, the goal of Spiritualism is to prove the existence of the spirit after death. First, mediums demonstrated to people that their loved ones still exist by relating accurate information, either through their own voice, automatic writing, or coded messages such as rapping. This evolved into physical mediumship which brought forth the voices of loved ones or materialized the likenesses of the spirits. Objects appeared in the rooms of some seances, and people experienced several types of physical phenomena.

Mediums did not deliver messages without question. Although Spiritualists were sincere about their beliefs, imposters and swindlers were not uncommon. To validate Spiritualism as a religion and mediums as authentic, groups of scientists and scholars tested mediums in a number of ways. Seances were held under controlled conditions, where rooms were checked for hidden objects and entrances, mediums even had their feet and hands tied to chairs to prevent sleight of hand.

James H. Hyslop was one of the early investigators. He was born in Xenia, Ohio in 1854 to Robert and Martha Hyslop. His twin sister died when they were only four months old. When Hyslop was ten, another sister and brother died of scarlet fever. The Hyslops were Presbyterians and James intended to enter the ministry when he was young, but while in college he developed doubts about religious faith.

Hyslop taught philosophy at Lake Forest University, Smith College, and Bucknell University in the 1880s. In 1891 he married Mary Fry Hall, an American he had met while in Germany. He took a position at Columbia University teaching ethics and logic. They had three children, two daughters and a son.

While at Columbia University, Hyslop wrote several textbooks and became involved with psychical research. After retiring due to ill health, he founded the American Institute for Scientific Research in 1904, planning to fund psychic research. He became a member of the Society for Psychical Research and later revived the organization as a section of his institute. He took on the role of chief investigator with Hereward Carrington and Walter F. Prince assisting. They published their first *Journal* in January 1907 and Hyslop was secretary-treasurer and director of the organization until 1920.

Hyslop published several books on psychical research, including *Science and a Future Life* (1905), *Enigmas of Psychic Research* (1906), *Borderland of Psychical Research* (1906), *Psychical Research and Survival* (1913), *Life After Death* (1918), and *Contact with the Other World* (1919).

Hyslop was convinced in the existence of an afterlife. He wrote in *Life After Death,* "I regard the existence of discarnate spirits as scientifically proved and I no longer refer to the skeptic as having any right to speak on the subject. Any man who does not accept the existence of discarnate spirits and the proof of it is either ignorant or a moral coward. I give him short shrift, and do not propose any longer to argue with him on the supposition that he knows anything about the subject."

After Hyslop died in 1920, his assistant and secretary, Gertrude O. Tubby, received communications from him through several mediums in the United States, France and Britain. "I find it difficult to assume that I am dead," he said through the medium, Mrs. Chenoweth. The messages were later published in her collection: *James H. Hyslop - X His Book: A Cross Reference Record* (1929).

SECTION III
ROOTS OF BELIEF

Chapter 8
Do Animals Go to Heaven?

Is heaven just an artifact of the creative human brain? A hallucination? A way to deal with the reality of death? Some believe primitive religion was a way to cope with death. Others think religion evolved as a mechanism to help bond communities when humans began to live in larger social groups. Early religions had no formal structure; the community shared mystical emotional experiences to bond its members. Did the concept of heaven emerge at the same time? Did it have an older origin in animals or are humans unique in their belief in an afterlife?

Self-awareness: Only for Humans?

Humans are more than just conscious; they are also self-aware. If consciousness is being aware of your body and your environment, then self-awareness is recognizing that you are conscious. In other words, to be conscious is to think; to be self-aware is to realize that you are thinking about thinking. Even in humans, self-awareness needs time to develop. Infants are conscious when they are born, but they are not yet self-aware. A classical test for self-awareness is to see if an individual recognizes themself in a mirror. It takes 18 to 24 months for children to pass the mirror test. Most scientists agree that self-awareness is the ability to see oneself clearly and objectively through reflection and introspection. But there are degrees of self-awareness.

If we define heaven as the existence of consciousness after death, one expects that conscious thought would be a prerequisite for survival after death. So, let's look at animals that have self-awareness to see how they deal with death.

On 7 July 2012, a group of neuroscientists convening at the University of Cambridge signed a document entitled "The Cambridge Declaration on Consciousness in Non-Human Animals," officially declaring that nonhuman animals, "including all mammals and birds, and many other creatures, including octopuses," are conscious. We have come a long way from the days when scientists considered all non-human animals as unconscious automatons.

Not only do we appreciate that animals are conscious, we have discovered that several species with more complex brains are also self-aware. Over the past thirty years, self-awareness has been reported in all the apes, elephants, bottlenose dolphins, killer whales, magpies and pigeons. Some also suspect that additional species may be self-aware, but our classical mirror test, mentioned in a previous chapter, might be biased toward sight as the source of self-recognition and needs to be updated.

Birds, those flamboyant creatures descended from dinosaurs, are one group that needs to be examined in more detail. Their brains are significantly different than mammalian brains, with no neocortex (the largest part of human brain that is responsible for our cognitive function), but there is evidence they possess near human-like levels of consciousness. Grey parrots, which have been studied extensively, show understanding similar to young children. They can identify colors, shapes and sizes, numbers up to six and use phrases such as "come here." Research on magpies has found that they respond in the mirror test like apes, dolphins, and elephants. Crows which have intelligence at least as high as monkeys, and maybe closer to apes, don't respond to the mirror test at all. Some bird species have also been found to have sleep patterns like mammals, including REM (Rapid Eye Movement) or dream sleep.

Pigs and dogs have been given the mirror test. Pigs react to their reflection and can use mirrors to find hidden food, but don't give clear responses that they are recognizing themselves. On the other hand, pigs can manipulate a joystick to move a cursor which researchers believe is a sign they can recognize actions as being made by themselves. Dogs don't respond to mirrors, and some believe their focus on scent makes smell a more important factor in recognition. A study in 2015 using a "sniff test" for self-recognition demonstrated that dogs do have self-awareness when it comes to identifying scents.

Animals and Death

Humans may be the smartest animal on our planet, but there are several species that approach our cognitive abilities, including chimps, dolphins and elephants. They have self-awareness, are highly social, can communicate in a variety of ways and some are even tool makers. Do they understand death? And if so, do they have any belief in existence of life after death?

In recent years, scientific studies have investigated responses to death in a variety of animals such as insects which practice corpse removal, burial, and cannibalism; birds that call and gather around the dead; and mammals that inspect dead bodies.

Some behavior seems to be protective. Mother dingoes will transport and care for a dead infant for several days as if they were still alive. Collared Peccaries react to the death of a herd member by pushing the body, staring at it, biting it, trying to pick it up, and defending it from predators. In one study, the herd visited and spent time with the dead body for 10 days after it died. Giraffes appear to react to their dead with guarding and carcass inspection.

Scrub Jays call each other when a body is discovered. One would expect this would be to warn others away in case there is a predator, but the calls summon others toward the body, instead of away from it. Crows also respond with calls and mobbing to notify kin that a member has died, and do not respond the same to the dead of other species of birds.

Chimpanzees

A chimp's brain is about one third the size of a human's brain, but they are one of the most intelligent non-human animals. They can learn to use picture tokens as well as sign language to communicate. In the wild, they modify sticks and leaves to create tools to gather termites and water. They possess culture, which is acquired information that is passed from one generation to the next. Do they also have some understanding of death?

Taï National Park is located near the Cavally River, which acts as a border between the Ivory Coast and Liberia. It is the largest rainforest in West Africa, with more than 200 native plants, 24 bird species and rare mammals, such as forest elephants, pangolins, pygmy hippos and Jentink duikers. It is also home to eleven primate species, including endangered western chimpanzees.

The park chimpanzees have been the subjects of behavioral research since 1979. The research group, headed by Christophe and Hedwige Boesch, has separate field camps to observe four communities totaling some 150 chimps. They have studied tool construction and use, cooperative hunting behavior, and many aspects of chimpanzee social life. They have also documented chimpanzee culture, previously thought to be only found in humans. But there is no indication that chimpanzees, one of our closest relatives, have any belief in the afterlife or heaven. Tina's death is an example of that.

It was 7:45 in the morning when research assistant, Gregoire Nohon, heard unusual chimpanzee calls in the forest nearby. Nohon, a female chimp named Bilou, and the alpha male, Macho, rushed to the site to see what was happening. When they reached the location, they found another dominant male, Brutus, circling the body of 10-year-old Tina. A leopard had attacked her. Brutus had probably frightened the predator away, but it was too late. Tina was dead.

Six males and six females gathered around the body. The males acted aggressively, dragging the body short distances. Brutus returned it to its place. Some of the females inspected Tina's wounds and the ground. Three males groomed the body for over an hour. They waved away flies and kept checking the body as if expecting that Tina might recover. In total, the group remained with the body for almost seven hours before abandoning it.

Numerous observations have been made of chimpanzees dealing with death. Infants who lost their mothers at Gombe National Park, Tanzania stopped playing and become lethargic as if in mourning. Mothers have been known to carry dead infants for long periods of time, sometimes until the corpse has rotted. But there is no ritual, no burial, and no indication that they believe life continues beyond the death of the body. Once an individual has died, their body is left and forgotten.

Dolphins

Although dolphin brains evolved differently to adapt to their marine environment and dolphins process auditory and visual information differently, their brain size is second only to humans in brain-to-body weight ratio. Like humans, dolphins can alter their surroundings, solve problems, and form complex social groups. They do not use tools, but they have a highly developed paralimbic system (integrates emotion, cognition and memory) which plays a large role in social and emotional bonds. Research has shown that dolphins have self-recognition, cultural learning, and can understand symbol-based communication and abstract concepts. But what about death?

Cetaceans' (whales, dolphins and porpoises) responses to dead members of their community have been documented over several decades. Females have been known to carry dead calves for up to a week. Several species of cetaceans have been shown to care for, attend to, be aroused by, or show interest in dead or dying individuals. A large study analyzed 78 records reported between 1970 and 2016, involving 20 of the 88 species. Toothed cetaceans, such as bottle-nosed dolphins and killer whales, were much more likely than baleen whales, such as humpbacks, to attend to the dead.

In one case, a dead female dolphin was found on seabed near the eastern coast of Mikura Island, Japan. Two adult males remained with the body, leaving only briefly to return to the surface to breathe. They stayed with the body until the third day, when the carcass disappeared. In another case, a dead young male was found wedged between two large boulders and was attended by at least twenty other dolphins, both male and female. As divers attempted to retrieve the body, groups of male dolphins displayed aggressive behaviors to keep the swimmers away. After divers finally retrieved the body, several of the dolphins followed the boat to port.

Cetaceans, like chimpanzees, respond to the dead with a variety of behaviors, but bodies are eventually abandoned.

Elephants

Elephants live in social groups and have very large and highly complex neo-cortexes. As a matter of fact, Asian elephants have the greatest volume of cerebral cortex of all existing land animals. Elephants form lasting relationships over decades and live in socially complex societies in which groups divide and merge over time. These complex relationships necessitate recognizing and remembering a wide range of individuals in their species. They are one of the few animals that can pass a mirror test. New research has also shown that they use sounds to identify individuals and give each other names.

In addition to self-awareness, elephants have complicated decision-making skills, both alone and with other members of their herd. They communicate with chirps and trumpets, nudges, kicks, and visual signals such as a tilt of the head. They can also communicate long distances with sounds that lie below the range of human hearing (infrasound). It is not surprising that matriarchs are able to identify at least 100 other individual elephants by their calls. But it is surprising that scientists have found that elephants can distinguish between three human languages they regularly hear: Maa (spoken by local herders), the language of the local Kamba people, and English (spoken by tourists).

Stories abound about elephant graveyards and herds visiting the bones of their dead. Is this myth true or just a tale? Scientists have been examining this question in South Africa.

Addo Elephant National Park is a wildlife conservation area situated close to Port Elizabeth. More than 600 elephants, 400 Cape buffaloes, over 40 endangered black rhinos as well as a variety of antelope species, lions, and spotted hyena live there. Nearby islands are breeding colonies for thousands of gannets and penguins.

Elephant researchers, Martin Meredith and Anthony Hall-Martin recorded the following death ritual in Addo. Eleanor was the matriarch of a family they called the First Ladies. Her closest associate was Maya. Eleanor gave birth to a calf about five months before her traumatic event. She suffered from a fall and had a swollen trunk, broken tusk and abrasions. When she fell to the ground around 6:00 pm, the matriarch of another family, Grace, tried to help her back up but was unsuccessful. Eleanor died about 11:00 am the next day.

When the body was first discovered, another female nudged it, tried to get it to move and stood over it, swaying. Over the next several days, it was visited by elephants that were part of Eleanor's family as well as those who were unrelated. They gently touched her body with their trunks and tried to lift her. After some time, the elephants were heard rumbling loudly. The calf wept and made sounds like a scream. Then the entire herd fell silent. They threw leaves and dirt over the body and broke off tree branches to cover Eleanor. The next two days, they stood quietly over her body, leaving only to get water or food. Over the course of several days, the carcass was visited by five other elephant groups, including several families unrelated to Eleanor.

Other research was conducted at the Samburu National Reserve in northern Kenya. Shifra Goldenberg and George Wittemyer recorded observations of elephants touching, approaching the dead individual and investigating the carcass. Behavior among the elephant differed. Some elephants made repeated visits while others did not.

Parveen Kaswan and Akashdeep Roy published a study in 2024 of five Asian elephant calf burials in the eastern Himalayan floodplains in West Bengal. Asian elephants less than one-year-old who died of natural causes were transported from forested areas to tea estates over long distances, one taking 48 hours. Calves were then placed on their back in irrigation ditches and covered with soil by several members of the herd. Unlike with African burials, where members return to the site to "mourn the dead," Asian elephants avoided returning to the site of the burials after their initial mourning vocalizations. This was the first recorded elephant burial.

With the use of infrasound, elephants can communicate over long distances. When a member of a herd dies, it's not surprising when distant individuals "hear the news" and arrive to see the body. But what about the death of a human?

Lawrence Anthony was an environmentalist, explorer and bestselling author. He was known for rescuing animals from the Baghdad Zoo during the Iraq invasion, protecting the Northern White Rhino in Sudan and creating Thula Thula, a 4500-hectare animal reserve in Zululand, South Africa. Unfortunately, the reserve ran into problems when the elephant's matriarch died, and the herd began to break through the fencing onto private land. Elephants were killed. Desperate, Anthony tried to communicate with them. Speaking to them at the edge of the reserve, he finally convinced the new matriarch to stay within the park boundary. From then on, Anthony became known as the "elephant whisperer."

Anthony died unexpectedly of a heart attack at the age of 61 in March 2012. According to his son, Dylan, they had not seen the herds close to the house for over a year and a half. The elephants had been staying at the far end of the reserve, a 12-hour journey. About a week after Anthony's death, the first herd arrived at their compound. A second herd appeared a day later. They stood around the house in an apparent vigil for two days, as if mourning Anthony's death. No one knows how the elephants knew about the elephant whisperer's death. It may seem magical but not only is their hearing more sensitive than ours, their sense of smell is also exceptional. What is unique is that when they discovered his death, they came to ritualize his passing.

We may never understand what happens in the minds of elephants, but they appear to mourn the loss of individuals, both kin and members outside their herd. There is some ritual involved with death, but like other intelligent species, there is no indication that they envision an afterlife. The belief in heaven must have evolved later, somewhere in the human family tree.

Chapter 9
Upright Explorers

We've examined animals' responses to death. For some species, there seem to be mourning periods and recognition of loss. For elephants, mourning rituals do occur. Burying a full-grown individual would be a major undertaking, but burials have been recorded for calves. Our closest relatives, apes, appear to distance themselves from the body after a short period of time. So, when did burial behavior start with humans? Is it a modern concept (within a few thousand years), or did some of our ancient ancestors mourn and bury their dead? Did they see death as the absolute end? Or did they envision an afterlife?

Let's examine the human family. We are *Homo sapiens*, but we still don't completely understand our ancient family tree. If you read the literature, it will seem like our ancestral story changes about every decade. That is because our view of it is changing. Scientists must make hypotheses about our ancestry from recovered skeletons and the remains of objects and tools that our ancestors have left behind. Thousands of fossils have been documented, but most are fragmentary, consisting of single bones or isolated teeth. It can be difficult to make educated guesses with only scattered remains. So, as new discoveries come in from around the world, the human story evolves, becoming more accurate. The ability to use DNA as a tool is also opening much discussion about our relationships with past ancestors. It is an exciting time of exploration into our past.

This is what we know, *at this time*, about the history of the hominin (human-like) family, and some of it is still controversial. *Homo habilis*, lived from about 2.5 to 1.5 million years ago. These people were only about four feet tall, walked upright, and used rock tools. That is why the species was nicknamed "handyman." The *H. habilis* people overlapped in time with several other Homo species in Africa and preceded *Homo erectus*, or "upright man." The *H. erectus* people stood about five feet tall. They not only used tools, but they also used fire and built shelters.

After *H. erectus* traveled the world, is the time period, from about seven-hundred thousand to one-hundred-twenty thousand years ago, that some scientists call the "muddle in the middle." It is a confusing time that includes different species of hominins in Europe and Africa, some of which are now being shuffled around to create new overarching species names. After that came our closest relatives, *Homo neanderthalensis* and *Homo denisovan*, who lived in Europe and Asia from about two-hundred to forty thousand years ago until modern humans became the primary hominin on the planet.

There are several death related behaviors which might enable us to see into the minds of some of our ancient ancestors. Primates and elephants carry or drag corpses, defend corpses and holding vigil near them. Unfortunately, fossils can't tell us about those practices. Burial is the most obvious sign of ritualized behavior and may indicate a belief in the afterlife. Did our ancestors bury bodies? And if they did bury the dead, what was their reason? If it was to keep predators away, we might find bodies buried far from living spaces. If burial had another purpose, like keeping a loved one close (maybe a type of ancestor worship), bodies might be found in living spaces. If the bodies were buried with possessions, that might be an indication that they needed the possession for their life in a world beyond death. Believe it or not, signs of cannibalism might even give us a clue to death rituals.

Artwork also provides a window into the consciousness of ancient peoples. Scientists once thought that using tools and language separated humans from animals. But we now know that chimps create tools out of sticks to gather termites and use leaves to make sponges. Gorillas and chimps can learn sign and symbol languages. Art has become a distinguishing human characteristic. It is born from our creativity, but what does it tell us about our views of the physical world and a belief in a heavenly afterlife? From simple cupules (small round depressions carved in rock), geometric markings, and the use of colored pigments to complex cave paintings and elaborate burials, art was an important component in the lives and deaths of our ancestors. Even without burial remains, art may also give us any indication about a belief in an afterworld.

In this chapter we will investigate the "Upright Explorers," the *Homo erectus* people. They lived through the Pleistocene Ice Age, from approximately 1.9 million until about one-hundred-forty thousand years ago. During this time, ice advanced and retreated in cycles. Mountain glaciers formed on all the continents and vast glaciers spread across North America and Eurasia. The African landscape also changed, from a land of forests to a drier, more open savanna. It was during these drier times that the *H. erectus* people evolved from their arboreal ancestors who preferred life in the trees.

Who doesn't like sitting around a campfire under a starry sky on a cool night? It's time to gather, to roast hotdogs or marshmallows, maybe make s'mores. As the fire crackles, it's not too difficult to imagine our ancient ancestors gathered around their own campfires. While hairier than modern humans, *H. erectus* people used fire. They were rather short, with a height averaging around five feet, walked upright like us and could have comfortably hiked for hours with no problem. Their pronounced brow ridges and low, flat skulls gave them a more ape-like appearance, but they were the first hominin species to have a projecting nose like ours.

H. erectus built simple dwellings, and created stone hand axes, knives and cleavers that could be used for hunting, butchering, and cutting. In Africa they killed big game, including antelope, rhinos, hippos and elephants. At one site, 50 types of fruits and vegetables were identified as being part of their diets.

What were they thinking while they warmed themselves by the fire? With a brain size about halfway between chimps and modern humans, they might have noticed the movement of the stars or moon across the sky or even thought about heaven. They had a hyoid bone in their neck which means they could have had a rudimentary speaking voice. They probably weren't singing around the fire, but they might have been acting out tales about a recent event.

As the first worldwide explorers in our family tree, *H. erectus* people adapted to new environments. From the open plains of Africa, they endured a Sahara Desert crossing to reach the colder north and trekked the tropical forests to the south. They crossed the open plains of Asia and the mountains and wetlands of China. Along the way they encountered antelopes, deer, big cats and wolves, giant beavers, large badgers, woolly rhinos, swine, giant turtles and even the largest primate on record, Gigantopithecus. They hunted everything from mammals to fish, turtles, crocodiles and amphibians. They were survivors in life, but what did they think about death? Did they believe in heaven or an afterlife? The presence of burials might give us an answer.

Burial

There has been no evidence found that *H. erectus* people buried their dead. The bones found don't have accompanying grave goods, and there is no indication they ritually used red ochre (a mineral used as a paint and later used in burials) either on or around any bones. There is some indication of cannibalism and mutilation in Java, Indonesia. The skulls found show cut marks suggesting that the skin of the head and brain were removed intentionally. Were they just eating the dead, or was something else happening? A death ritual perhaps?

Most people react negatively to cannibalism, which is not surprising, but it has been a widespread custom found on most continents going way back into prehistory. Some rituals even lasted into modern times in parts of West and Central Africa, Melanesia, Australia, New Zealand, Polynesia and the Americas. For some societies, flesh was regarded as food. For others, eating a person's flesh was used in rituals to obtain magical powers or the person's vitality. The Aztecs practiced cannibalism as part of the ritual religious sacrifice. The Māoris feasted on the flesh of their enemies. Some ate the bodies of their relatives as an act of respect.

H. erectus people may have mourned their dead, just not the way in which we are accustomed. Because of the limited number of fossils recovered, we undoubtedly don't have a very clear picture of what was going on at the time. Were their burials in shallow graves that eroded over time? Maybe they cremated their dead, de-fleshed and burned the bones in some sort of ritual, or practiced tree burials like the Sioux, Ute, and Navajo. Maybe we just haven't dug the right location for a burial site. Only time will tell.

Symbolic Art

Think about all the art that is associated with modern religion, burial and belief in the afterlife. Every religion has their own symbols: Christian Cross, Star of David, Menorah, Baha'i Nine-pointed star, Om, Shinto Torii, Sikhism Khanda, Taoism Taijitu, Islam Star and Crescent, and Dharmachakra, to name just a few. There are deities, angels and other related beings represented in various kinds of artwork. Items associated with burials include special clothing, coffins, and burial shrouds. Tombs and tomb stones are also created.

It appears that the question of burial is still open for the *H. erectus* people, however, they appear to have been creating some art. Was their art just for decoration or did it have some meaning? This is what we know so far. They etched designs on objects and in caves. They were using ochre, a pigment that was later used in burials of other hominins. And they created some of the first crude mysterious Venus figures.

One of their earliest artistic carvings was found at Trinil on the banks of the Bengawan Solo River in Ngawi Regency, East Java Province, Indonesia. In 1891, Dutch anatomist Eugène Dubois discovered what he called "Java Man," now confirmed to be *H. erectus*. It wasn't until recently that researchers began to study 166 shells recovered from the site.

A couple of the shells are noteworthy. One has a smooth and polished edge, suggesting it may have been used as a tool for cutting or scraping. The second shell held a bigger surprise. It is engraved with an M-shaped zigzag design, likely carved with a sharp object, such as a shark tooth. It's unclear what the design means. It may have been a decoration, or a practice run for decoration on another object. Maybe it had some spiritual significance.

In 1969, in Bilzingsleben, Germany, archaeologists discovered engravings on an extinct forest elephant bone. The fragment of tibia contained groups of parallel lines with V-shaped cross sections created with a single stone tool. Remains of the *H. erectus* skulls which had been intentionally smashed after death were also found, maybe as part of a burial rite. Similar markings appear on a forest elephant vertebra at Stranska Skala in the Czech Republic. No one is sure what the lines mean. They might be maps, a calendar, or maybe a type of early writing.

Ochre pigment is an iron-rich oxide can also be used to create red-yellow-brown paints and dyes for a wide range of uses, including rock art, pottery, wall paintings and human tattoos. Ochre is also important because it is found with human burials. Archeologists didn't think the *H. erectus* people were using the pigment but then came the surprise. The Kapthurin Formation in Kenya is in the Rift Valley of east Africa. It is the setting of the earliest use of ochre discovered at a *H. erectus* site called GnJh-03. A total of five kilograms (11 pounds) was discovered. Why is ochre so important? *H. erectus* might have been using the ochre for practical uses, but other discoveries hint that *H. erectus* people might have been up to more than that.

Cupules are the earliest known type of prehistoric art. These small, rounded pits or hollows carved into rock surfaces have been found across the world. The oldest cupule-bearing rock is a rounded cobble discovered in the primordial Olduvai Gorge in Tanzania, dating to approximately 1.7 million years ago. Those were probably utilitarian hollows created by grinding rather than art, but other cupules have been found since.

Archaeological investigations began at the Bhimbetka complex in India in the 1970s. The oldest cupules with symbolic purpose were discovered during the 1990s in two quartzite caves: Auditorium

Cave at Bhimbetka and the Daraki-Chattan rock shelter. The quartzite rock formations at Bhimbetka contain over 700 rock shelters, with over 500 with cave murals and other forms of more recent art. The big surprise was that some of the cupules were hammered out of the rock surface by the *H. erectus* people. Daraki-Chattan Cave in the Indragarh Hill near Bhanpura, central India, features more than 500 cupules on its walls, floor and rock fragments in the sediment. Not only that, a small group of cupules on the south wall contained elements of red ochre pigment in them.

In addition to cupule carving, the *H. erectus* people may have been the first to carve rudimentary Venus figures. Venus figures are small statues resembling women. The name was coined by Marquis de Vibraye who discovered one of the first figures in ivory and named it after the Roman goddess Venus. Most of the figures have wide hips and legs that taper to a point. Arms and feet are usually absent, and the head is usually small and faceless. Some statues have exaggerated hips, breasts, abdomen, thighs, or vulva. Most have been unearthed in Europe and were made by more modern humas.

Scholars have differing opinions on the purpose of the figurines. It has been suggested that they served as a ritual or symbolic function. They could have been religious figures, an expression of fertility, goddesses, or as self-depictions by female artists. No matter what their purpose, they do seem to have some symbolic function, and they were definitely not expected to be found with the remains of *H. erectus* people.

The Berekhat Ram Venus was excavated from the Golan Heights site in 1981 and first described by Naama Goren-Inbarof the Institute of Archaeology, The Hebrew University of Jerusalem. The stone's natural shape resembles a woman, but the story gets interesting when it is examined under the microscope. Grooves were carved with a sharp-edged tool to create a neck and arms. The stone was also scraped and smoothed in some areas to give it more of a female shape. A second *H. erectus* Venus was found in Tan-Tan, Morocco in 1999. The shape of the stone resembled a human figure with arms and legs but was thought to be entirely natural. Again, a microscopic examination suggested the shape was altered to accentuate the body. It also contained traces of iron and manganese pigment suggesting it had been painted.

The *H. erectus* people didn't leave many clues about their beliefs surrounding death or the possibility of an afterlife. Did they not bury their dead or have we just not found a burial? The clues they left behind are intriguing. What was the purpose of the 500 plus cupules carved in the cave. Were they keeping track of time, maybe marking the passing of the full moon. Were they recording the deaths of loved ones? And what about the Venus figurines? Was it a symbol of an earth goddess? Scientists are still puzzled about Venus figurines, but we know they continued into more modern times and were made by *H. sapiens.* It appears the *H. erectus* people will remain a mystery, for a while at least.

Chapter 10
Archaic Peoples

Several subspecies of *Homo* can be grouped into the broad category called "archaic humans" in the period that predates the earliest modern humans about three-hundred thousand years ago. These include *H. ergaster, H. antecessor, H. habilis, and H. rhodesiensis.* They had brain sizes almost as large as modern humans, thick skulls, prominent brow ridges and lack of a prominent chin.

In this book, I'll use as an example the *Homo heidelbergensis* people (who may have another name soon) who lived from about seven-hundred to four-hundred thousand years ago in both Africa and Europe. The species gets its name from a 1908 discovery in a sandpit near Heidelberg, Germany. A workman found a mandible that was nearly complete, except for a couple of missing teeth. Since the first fossil find in Germany, many other fossils have been discovered.

Like H. erectus, *H. heidelbergensis* people were explorers, traveling to many parts of the world. They were about five and half feet tall, stocky, with brains just a bit smaller than our own and large brow ridges and flatter faces. The oldest site is in Bobo, Ethiopia (about six-hundred thousand years ago). There are numerous sites in Europe, one of the most interesting containing about 29 individuals found in the "Sima de los Huesos" or the "Pit of Bones" site in the Atapuerca Mountains of Spain from about three-hundred thousand years ago. There are also sites in Asia and Africa.

While their ancestors may have sat around their campfires at night, *H. heidelbergensis* people, particularly those who had moved out of Africa, had ice ages to contend with. They needed fire, shelter, clothing, and newer tools to survive. and appear to be the first humans to live in colder climates and hunt large animals such as deer, horses, elephants, hippos and rhinos. They were also the first to build shelters out of wood and rock and use hearths to control fire.

Burials

H. heidelbergensis people had advanced in many ways from their predecessors, but did they believe in an afterlife? Imagine that you have just lost a loved one, a beloved grandmother, a sister or maybe a husband struck down during a hunt. The body that once housed the living spirit of the person is all that is left. You have seen death before; it is a natural cycle of life. Mother earth gives this life and in the end the body must be returned to her. The people of your small clan gather and carry the body to a sacred cave, a place where your loved one is returned to the mother's womb. Torches are lit. The way inside the mother is long, slippery and cold. Sounds echo and the walls seem to close in around you. Finally, you reach a black pit.

The account above is more than just a story. Sima de los Huesos (The Pit of Bones) is an exciting *H. heidelbergensis* find, not only because it contains many individual fossils of all ages, but because it looks like a burial pit. The pit is located deep inside the Cueva Mayor-Cueva del Silo cave system along the southern edge of Spain that boarders the Mediterranean Sea. The pit sits a quarter-mile's walk from the entrance of the cave, deep within a network of dark tunnels that include other older fossil sites.

From the cave floor, the pit drops off to become a slanted shoot that descends about 40 feet. Scientists considered several hypotheses to explain how the bodies got to the location. There is no evidence, like the remains of fires for cooking, that *H. heidelbergensis* lived in the cave. There are no bite marks indicating that predators carried the bodies to the location. There is no sign of flooding, so they didn't appear to be victims of some catastrophic event like a cave-in. The individuals appeared to be already dead and intentionally deposited into the cave by others over a period of a few years.

We can't know what the *H. heidelbergensis* people were thinking as they added bodies to the pit. There are no remnants of any type of physical ritual, even though one hand axe was found. The six-inch-long blade might not appear to be important to us, but it was a carefully crafted tool made of high-quality quartzite which was rarely used in that region. Would such a possession be wasted on a burial if it didn't have any significance? Were they just getting the bodies out of the way or were they sending them on their way to an underworld? Maybe future discoveries to give us more clues.

Symbolic Art

None of the bodies found in Sima de los Huesos had been covered with body paint, but we know that some *H. heidelbergensis* people were using pigments at that time. In Pinnacle Point Cave, a small promontory immediately south of Mossel Bay on the southern coast of South Africa, Cave site PP13B, ochre was found, presumably for body painting. It appears that art and burial were two separate activities for the *H. Heidelbergensis* people. If they thought their dead had souls or that those souls were moving on to another life after death, they did not decorate their bodies or send them with any grave goods for the next world.

Chapter 11
People of the Ice Ages

As I have mentioned before, our understanding of the human family tree has changed over time as we learn more about our ancestors through fossil evidence and DNA. *Homo neanderthalensis* (Neanderthal) people appear to be the descendants of the European *H. heidelbergensis* ancestors. At the same time, modern *Homo sapiens* people appear to have evolved from other populations of *H. heidelbergensis* in Africa. There is also a sub-population of Neanderthals named Denisovans recently discovered in a region in Russia. On top of that, there was crossbreeding going on and present-day human populations still carry Neanderthal genes. I know, it can be confusing. Thankfully, we are only discussing burial practices and art in this book.

The name Neanderthal comes from the Neander Valley in Germany where the first remains were discovered in 1856 in Feldhofer Cave near Dusseldorf. Neanderthals are considered as a distinct sub-species of humans that emerged about two-hundred thousand years ago. They lived in Europe and spread eastward through the Mediterranean and Asia. Because they lived in an area of abundant limestone caves, their bones were well preserved when left or buried in caves, and they are better known than many more ancient species that lived in open areas.

Except for modern Inuit people, Neanderthals lived in some of the coldest and most challenging environments on the planet. They inhabited Europe during a glacial period followed by a warming period, and then a second glacial period. The cold climate supported mammals like mammoths, mastodons, woolly rhinos, and saber-toothed cats. They were a good source of food, if they could be killed.

Neanderthals were shorter and stockier than modern humans, but their brains were a bit larger. Some defining features of their skulls include the large brow ridge and wide face, angled cheek bones, and a wide nose for humidifying and warming the cold, dry air.

Neanderthals used a diverse set of sophisticated wood and stone tools, controlled fire, lived in shelters, made and wore clothing, and were skilled hunters and gatherers. They also made some symbolic or ornamental objects. Until the early 2000s, it was widely thought that Neanderthals lacked the capacity for speech, but recent studies have found that Neanderthal vocal physiology is like modern humans. Language was entirely possible.

Burials

Neanderthal fossils of well over 200 adults and 70 juveniles have been collected over the last century. Sites have been found in southern France, the northern Balkans, Syria. Iraq, Ukraine and Russia. There is evidence that Neanderthals deliberately buried their dead and may have even marked their graves with offerings, such as flowers. Two of the oldest sites may have included cannibalism, so we will begin there.

Neanderthals may have had good reason to consume the flesh of others during bad times when little food was available. But there are also clues that removing meat from the bones before burial may have been a ritual as well. Were Neanderthals doing more than just burying the dead? Did they envision souls and afterlife? Here is what we know so far.

We will start with Krapina, a town in northern Croatia, in Krapina-Zagorje County, a hilly region about 30 miles from Zagreb. In 1899, on a hill called Husnjakovo, archaeologist Dragutin Gorjanovic Kramberger discovered fossils belonging to about 80 Neanderthal individuals between the ages of 2 and 27 years. They were found in shallow graves along with stone tools and animal bones that were deposited over a ten-thousand-year period.

Some marks resemble the marks of cutting meat from the bones of game animals, suggesting they might have been practicing cannibalism. Others believed that carnivores were responsible for much of the bone damage. But other cut marks are unexplained. The bones could have been de-fleshed to prepare for the burial.

The most complete and intriguing skull belonged to a 20-year-old Neanderthal woman. There were 42 cut marks made across her head after death. The markings do not indicate brain removal or even butchery of the large chewing muscles. So, what was going on? David W. Frayer and colleagues concluded that the parallel marks running across the forehead were a sign of some sort of ritual behavior. Davorka Radovčić, a Neandertal specialist and curator of the Krapina collection, also suggested they may have been practicing burial rituals.

In 1908, a suspected burial was found by three brothers in La Chapelle aux Saints, a small hamlet in the rolling hills of southwest France about halfway between Bordeaux and Lyon. The discovery did not come without controversy. Some thought the Neanderthal had just died in the cave, while others considered it a purposeful burial. It wasn't until almost 100 years later that scientists revisited the site and confirmed that it was a burial. Another site was found at La Ferrassie is not far from La Chapelle aux Saints, located in Savignac de Miremont in the Vezere valley. It is a deep cave flanked by two rock shelters. The cave was found to hold a total of eight Neanderthal individuals, including adults, children, infants, and two fetuses. Like La Chapelle aux Saints, the burials were in question. Researchers at the CNRS and the Muséum national d'histoire naturelle (France) and the University of the Basque Country (Spain) used a variety of criteria to prove that at least the one child was buried purposefully. The skeleton had been purposefully positioned with its head to the east, higher than the pelvis.

A third burial site made the headlines across the world. Shanidar Cave is located on Bradost Mountain in the Zagros Mountains, in the Kurdistan Region of northern Iraq. This site held the remains of 10 Neanderthal men, women and children buried about fifty thousand years ago. One was found with ancient pollen clumps, and scientists thought that Neanderthals may have included flowers as part of their burial. However, it has been suggested that the pollen was introduced to the burial by a gerbil-like rodent. The Persian jird stores large numbers of seeds and flowers in their burrows which may have crossed into the site. So, flowers placed directly in the grave may not have been the case. But more discoveries were yet to come.

In 2014, more researchers arrived at Shanidar to re-excavate and found additional bones near the "flower grave." They unearthed another male in his 40s with a damaged but complete skull and chest, plus both forearms and hands. The remains were analyzed using a variety of techniques that were not available when the cave was first investigated. Interestingly, the male looked like he was placed on his back, with the shoulders and head raised, and the head resting on its left side on top of the left hand. There was also a triangular stone which had been placed under his head and right shoulder like a pillow at the time of burial. The researchers found plant material in the rock immediately above the bones.

In another cave in southwest France, La Ferrassie, scientists discovered an incredible grave. The site is situated in the Vezere valley in deep cave flanked by two rock shelters within the limestone cliff. Excavation began at the site during the early part of the 20th century, under the supervision of Louis Capitan and Denis Peyrony. Fossils, stone tools, and other signs of occupation were found to belong to Neanderthals. One of the most interesting findings was a large limestone slab covering the grave of a Neanderthal child. On its underside was an arrangement of cupule-art, consisting of two large and eight pairs of smaller cupules. In a previous chapter, I discussed cupules created by *H. erectus*. Now we see them in a Neanderthal grave. Is this just a coincidence or are cupules an important symbol?

Neanderthals apparently buried their dead, but the purpose of burial is unclear. La Ferrassie Cave gives us some tantalizing clues about burial and cupule art. But we can only guess its purpose. For *Homo sapiens*, burials were much more involved, with pigment-covered bodies and grave goods. We will look at some of those cases in the next chapter.

Symbolic Art

Although we haven't found any evidence of Neanderthals using pigments in grave sites, they were using them for body painting as well as wearing feathers and claws of raptors and corvids and shell beads with pigment. They were also painting symbols on cave walls.

Several important sites hint to what was happening at the time. At the Maastricht-Belvèdère site in the Netherlands, red hematite imported from dozens of miles away was being used. Pigment was found in Antón, a large rock shelter by the Mula river and Cueva de los Aviones, a large cave near the shore of the Mediterranean Sea. Red ochre pigment was also used to decorate shiny white stalagmites in the Caves of Ardales, near Malaga in southern Spain around sixty-five thousand years ago. The purpose of the art is unknown, but our ancient cousins were intentionally modifying the cave for a specific purpose, maybe for mystical or spiritual reasons.

Pigments can also be used to create cave wall art. The Cave of Maltravieso in west central Spain was discovered in 1951, when an explosion inside a limestone quarry revealed the deep cavity. Inside were human skulls, animal remains, and ceramics. There are three hand stencils made with pigments on one wall made by Neanderthals.

Cueva de La Pasiegai located in the Spanish municipality of Puente Viesgo is an example of a cave that housed first Neanderthals and then *Homo sapiens.* When it was first explored, discoverers were impressed by the more than 700 painted forms, including: 97 deer, 80 horses, 32 ibex, 31 cattle, two reindeer, as well as a bird and a fish created by *Homo sapiens.* What they originally missed was much older, simpler artwork created by Neanderthals. These included more than 130 varied symbols including a ladder shape, red dots and other figures.

Neanderthal carving was also found in a cave that was occupied by both Neanderthals and *H. sapiens.* The Die Einhornhohle (Unicorn cave) cave is in the Harz mountains in Central Germany, 241 kilometers (150 miles) southwest of Berlin. Neanderthals left behind a carved and decorated deer toe bone fifty-one thousand years ago. The front of the bone is carved with lines in the shape of inverted V's that appear to point in an upwards direction, with smaller incisions on its lower edge, which was probably its base. The carved bone was found alongside the shoulder blade bones of deer and the intact skull of a cave bear.

Another interesting carving was found in a Mediterranean seaside cave in Gibraltar called Gorham's Cave. Researchers discovered X-shaped markings etched into a stone on the floor. The scientists have no idea what the markings mean. Is it abstract art, some sort of map, or maybe a message? We may never know.

Pigment use and etchings by Neanderthals are altering our views of our ancient cousin, but nothing could have been more shocking than a find located in Bruniquel Cave in an area east of Montauban in southwestern France. For all intents and purposes, the structure looks like mini-Stonehenge built by Neanderthals around one-hundred-eighty thousand years ago. Researchers discovered two large ring-shaped stalagmite formations and four smaller arrangements 336 meters inside the cave. The structures were made from hundreds of pillar-shaped stalagmites, which were chopped to a similar length and laid out in two oval patterns up to 40 centimeters (16 inches) high. The stalagmite pieces are also covered with soot, indicating fires were burnt there. The Neanderthals who built them had to travel deep into a cave where there was no natural light. They would have cooperated to build the rings, using fire to illuminate the cave. We have no idea what the rings were for.

It appears that Neanderthals were burying their dead and may have been holding rituals in association with the death and burial, but did they believe in an afterlife? We know that in more modern societies that believe in an afterlife, bodies are often accompanied by objects from life. We haven't really seen that in Neanderthals, but they appear to have been creating symbolic art. It isn't clear what their belief in the afterlife was.

Chapter 12
The Wise Humans

Both fossils and genetic evidence indicate that Neanderthals and modern humans evolved from a common ancestor between seven-hundred and three-hundred thousand years ago. While Neanderthals were living in Europe, *H. sapiens* ancestors remained in Africa, not migrating north to the middle east and then Europe until about forty thousand years ago. Unlike the questionable graves of other ancient people, it is obvious that our direct ancestors were burying their dead and many times including burial items. If burial items are a sign that they believe in an afterlife, they may have held this belief as early as one hundred thousand years ago.

Burials

Most ancient *H. sapiens'* burials were plain and included items of daily life. They placed bodies into pits and included ornaments worn while alive. A study of 85 burials showed that men were found buried more often than women and infants, who were buried sporadically. There were also a few ornate burials in Russia, Italy and the Czech Republic dating back nearly thirty thousand years.

In Skhul Cave, part of the Valley of Caves, the remains of seven adults and three children were discovered in the cave from 1929 and 1935. These individuals had both Neanderthal and human characteristics. At least a few of the individuals were buried intentionally. This is one of the earliest graves where the deceased was buried with an object. The mandible of a wild boar was placed on the person's chest at the time of burial.

A second site, Qafzeh cave, is located south of Nazareth and was first excavated by R. Neuville and M. Stekelis in the 1930s, and again between 1965 and 1979 by Ofer Bar-Yosef and Bernard Vandermeersch. More than 25 skeletons have been recovered. There is evidence of the deliberate burial and positioning of several bodies in the cave. *H. sapiens* people also used ochre pigment to paint the body and marine shells for ornamentation. One juvenile who had suffered a brain injury was buried with a deer skull and antlers placed in his left hand.

Even though we know that modern humans evolved in Africa, few graves have been found there. That may be because less research has been conducted in the African continent, but that is changing. One new burial found in Panga ya Saidi Cave in Kenya was less than 10 miles inland from the southeast coast and contained the remains of a toddler buried about seventy-eight thousand years ago. The child, named Mtoto by the researchers, was buried in a pit purposefully dug for the grave and covered with earth. The skull base and attached neck bones had sagged over the years suggesting that its head originally lay on a support or pillow made of fibrous material that had decayed. Rotation of a collarbone and two ribs suggested that the upper body had originally been wrapped in a shroud.

One of the most elaborate ancient burials was found in Sunghir, located north-east of Vladimir, Russia, adjacent to the Sunghir Stream and Klyazma River. The burial grounds were accidentally unearthed in 1955 while the area was being quarried. Once work was halted, careful excavations from 1957 to 1977 led to the discovery of many remains. About thirty-five thousand years ago, hunter-gatherers who roamed plains buried their dead there. At least 10 individuals were found. But nothing could compare to the body of a middle-aged male who was covered in ochre. In addition, his grave contained about 3,000 mammoth ivory beads, pierced fox canines, and ivory armbands. The grave of a juvenile and adolescent, buried head-to-head, also contained beads and ochre, mammoth ivory spears, ivory disks, and pierced cervid antlers.

Ancient humans were travelers like other members of the *Homo* family, moving across southeast Asia and into Australia. The oldest human fossil remains found in Australia date to around fifty thousand years ago. The people living there had much larger bodies and more robust skeletons than they do today. They also cremated and buried their dead.

Lake Mungo is located in south-west New South Wales, Australia. Mungo Woman was discovered in 1968. At forty-two thousand years old, her remains are the earliest ritually cremated remains found anywhere in the world. Mungo Man was discovered in 1974. He was laid out on his back for burial and covered in red ochre before being buried in the beach sands along the lake. Kow Swamp in northern Victoria was excavated between 1968 and 1972. The human skeletons were dated to about ten thousand years ago. The remains of over forty individuals were found and included men, women and children. This burial site is one of the largest from this time period anywhere in the world.

It is obvious that by the time the *H. sapiens* people began their trek out of Africa, they were burying their dead. Did they believe that the souls of those dead were passing on to another life in a spirit world?

Burial with artifacts may be an indication that they thought people needed objects in the next world, but we can't be sure. Their artwork may offer some clues.

Symbolic Art

In addition to burials, the first examples of *H. sapiens* art date from between one-hundred to sixty thousand years ago in Africa. Our ancient ancestors left artwork behind in the form of carvings, jewelry created with natural objects, items colored with minerals like red ochre, and probably the most impressive, cave art. Does their art give us a clue to what they were thinking about the afterlife? We will look at just a few examples of art they left behind and what it might mean.

Blombos Cave is in Blombos Private Nature Reserve, east of Cape Town, South Africa, and contains artifacts dating back to one hundred thousand years. The only human remains found were a few teeth and they appear to be modern *H. sapiens.* Of the more than 8,500 fragments of ochre dating from the middle stone age, 15 fragments showed evidence of engraving. They contained geometric crosshatch-incised patterning. Geometric designs were also observed on an engraved bone fragment. The marks were intentionally engraved and there was also evidence of bone working techniques at the site. Blombos type engravings are not isolated occurrences. They were also discovered in a half dozen other sites, including Klasies River, Diepkloof Rock Shelter, Klein Kliphuis and Wonderwerk Cave, implying that art and symbolic traditions were shared across southern Africa. The patterns may have meaning, but we don't understand them at this time.

About twenty-five thousand years ago, in a rock shelter in the Huns Mountains of Namibia on the southwest coast of Africa, a *H. sapiens* drew an animal in charcoal on a hand-sized slab of stone. The stone was left behind and became buried on the floor of the cave by layers of sediment. It wasn't until 1969, when a team led by German archaeologist W. E. Wendt excavated the rock shelter, that the stone was found. Between 1969 and 1972, Wendt unearthed several painted slabs in a cave he named Apollo 11, after NASA's successful moon landing mission at the time. There were seven painted brown-grey quartzite stone slabs in all, depicting a variety of animals painted in charcoal, ochre and white pigment. The animals are not easily identified but might be felines or bovids and perhaps a zebra, giraffe or ostrich. It's interesting that this early art is of animals. Thousands of years later we will see something similar in the caves of Europe.

By forty thousand years ago, cave art was being produced by *H. sapiens* in Europe and Australia. Most of the art depicted animals and/or probable spiritual beings, but smaller marks in many caves in France, and possibly others in Europe, are now being analyzed. New research suggests they may be a written 'code' familiar to many prehistoric tribes. Twenty-six symbols appear repeatedly across thousands of years, some of them in pairs and groups in what could be a rudimentary language. These suggest that early *H. sapiens* were attempting to represent ideas symbolically rather than realistically and share information across generations. Two of the symbols can be seen in Celtic tradition, a spiraling circle and tree shape. For the Celts, the spiral represents the connectivity of everything and the constant growth of life. The tree is a symbol used for Beltane, the celebration of summer.

Venus figurines, much more refined than those found with *H. erectus,* were also widespread in Europe. An ivory female head with bun from Dolni Vestonice, Czech Republic, is one of only 2 human head carvings from this period that show eye sockets, eyelids and eyeballs. It is dated to twenty-six thousand years old. What they represent is still a mystery.

Europe is also home to spectacular cave art. Nearly 350 caves have been discovered in France and Spain alone. The earliest known paintings of figures are found in Coliboaia Cave and Chauvet Cave in France dating to around thirty thousand years ago. The most common subjects were large wild animals, such as deer, bison, and horses. There were also tracings of human hands and abstract patterns. Drawings of humans were rare and are usually schematic as opposed to the more detailed and naturalistic images of animal subjects.

In Southwest Asia, we have found paintings that depict several human figures hunting pigs in the caves in South Sulawesi, Indonesia.

In Australia, rock art was expressed in painted and carved forms and still plays a significant role in Aboriginal culture today. It's estimated that there are more than 100,000 significant rock art sites around Australia containing petroglyphs (rock engravings) and pictographs (drawings).

Rock art in Australia employs two main design types. The first uses engraved geometric forms, such as circles, concentric circles, arcs, dots or animal tracks. The second creates figurative forms, such as silhouettes of humans or animals. These figures can be either simple outlines or more complex drawings. Some even show the internal organs of humans and animals. Stories about these artworks have been passed down, in some cases, through thousands of generations. We are fortunate that Aboriginal people are still here today to tell the story of their art.

Ancient people brought art with them to North America. A fossil bone unearthed in Florida was engraved with the figure of a mammoth around thirteen thousand years ago. The Winnemucca Lake petroglyph carvings near Reno, Nevada are large and complex. Deeply carved lines and cupules form complex designs. The carvings are commonly 1–2 cm deep with cupules occurring on vertical versus horizontal surfaces. Some of the petroglyphs may represent clouds and lightning, others are diamond- shaped, and there are some patterns that might represent trees. There are no animals, humans or handprints that are found at more recent sites. The Pyramid Lake Paiute Tribe, which owns the site, considers them sacred to this day. The earliest petroglyph in South America is an anthropomorphic figure in central-eastern Brazil dating back twelve thousand years.

What is the explanation for all this art? Hunting magic has been suggested. Some believe that in times when it was difficult to find game, the artists could increase the fertility of the herds or help the hunter be successful by painting the images of success beforehand. Early archeologists thought this could have led to religious practices that we see today. The problem is, there is little correlation between the animals painted in the caves and the actual remains of animals that were being eaten by the early people.

Another explanation was that early people were creating art for art's sake. There are two problems with this idea. First is the location of most of the paintings. If one is creating art to share, then why is the art located in such inaccessible locations. The second problem is that creating this art was an expensive endeavor. It was taking up time that could have been used on other jobs such as hunting or gathering. It also put one's life at risk while traversing slippery cave passages and maybe running into wild cave bears or big cats.

A third explanation, which has been discussed in detail by David S. Whitley, is shamanism. A trance state can bring forward mental images. During the first stages of trance, the imagery is dominated by geometric light patterns. In the second phase, the original pattern is influenced by personal and cultural influences. Finally, in the third phase, there are full-blown hallucinations. The person may also imagine becoming what he or she hallucinates.

Whitley wrote, "As the 'entrails of the underworld,' they were the vortex that, through ritual trance, the shaman used to access the supernatural. And the fact that sensory deprivation, resulting from isolation in the deep caverns, can easily lead to an altered state of consciousness...."

If achieving a trance state, and altered state of consciousness, is an explanation for what's inspiring *H. sapiens* cave artists, then what about the earlier art of Neanderthals and *H. erectus*? Were they also able to alter their conscious thoughts? Were their visions just hallucinations, or was something else happening? We may never know the answer, but we can follow *H. sapiens* in their journey as inspired art gave way to megaliths, ancient spiritual beliefs, the first religions and numerous descriptions of heaven.

Chapter 13
Megaliths

Human belief in an afterlife and altering consciousness did not jump straight from burials and cave paintings thirty thousand years ago to the *Egyptian Book of the Dead* written about thirty-five hundred years ago. Ancient people were not writing their thoughts down, but they did leave massive stone structures and artwork that give us clues to what they were thinking. Some of those structures not only give us clues about ancient people's artistic and spiritual traditions, they suggest that people were looking at the sun, moon and stars, at least to predict seasonal changes.

Around fifteen thousand years ago, the Earth started warming. Large ice sheets, which covered significant parts of North America and Europe melted, leaving behind land rich for grazing, gathering wild grains and eventually agriculture. This warming trend was briefly interrupted by a cool period called the Younger Dryas between roughly 12,900 and 11,600 years ago which returned parts of the northern hemisphere to ice age conditions for a time.

People survived these harsh conditions, living in caves for shelter and using fire to create warmth. As the planet rewarmed, average temperatures increased to near present-day levels. Anatolia, the Asian part of Turkey, was a hotbed of development after the Younger Dryas. It was likely forested and capable of sustaining a variety of wildlife, plants and grains. During this time, many villages were built (about 40 prehistoric sites have been found), including one of the most interesting *H. sapiens* sites unearthed in a century, Gobekli Tepe.

Gobekli Tepe and Çatalhöyük

Six miles from Urfa in southeastern Turkey, Klaus Schmidt was intrigued when he came across the tops of carved stones buried in a hill called Gobekli Tepe (Potbelly Hill), named for its mound-like appearance. It seems that local farmers had walked right by these stones for hundreds of years without a second thought. They were busy farming, and the artifacts were just stones sticking out of the ground.

As Schmidt carefully unearthed the site, eleven-thousand-year-old megalithic stones were the first objects uncovered. These carved stones were arranged in a circular structure by people who had not yet developed metal tools or even pottery. Since then, additional buildings that functioned as homes, quarries, and stone-cut cisterns have also been unearthed. People also buried the bones of their dead in the floors of the homes, maybe to keep the spirits close.

The most impressive buildings are large circular structures, roughly 98-feet (30 meters) across, supported by massive t-shaped stone pillars. The pillars are decorated with abstract pictograms and animals carved in relief. The mammals include lions, bulls, boars, foxes, gazelle, and donkeys, reminiscent of older cave paintings. There are also vultures which may be linked to burial rituals. The large, central pillars have human arms and loin cloths, suggesting they were stylized humans (or perhaps deities) with no heads. It is believed that these buildings had wooden roofs with central holes to descend into the chambers.

Schmidt proposed that the people were engaged in shamanic practices. Imagine living in such a village. Your days might be spent hunting wild cattle, gathering local grains and plants, making stone tools and clothing, building dwellings, or maintaining a home for your children. There are no caves here for a shaman to paint magnificent beasts, but the people are well fed and have enough free time to create a cavern of their own. They carve not only pillars to build the cave, but they also engrave important beasts on those pillars. When the shaman calls the people inside, they climb down the latter into the dark womb of Mother Earth. Torch light wavers in the darkness and the animals come to life.

Manipulated skulls were also found at Gobekli Tepe. In 2017, researchers revealed that human skulls may have once been on display. Marks on the three partial skulls indicate that they were de-fleshed, painted, and could have been suspended by a cord that passed through a small drill hole at the top. Julia Gresky of the German Archaeological Institute said, "The three modified skulls attest to the special treatment of certain individuals and represent an entirely new category of find, one which testifies to the interaction of the living with the dead at this important Early Neolithic ritual center."

The Vulture Stone is one of the more complex reliefs excavated at the site. One vulture holds the human's head in its wing. Müslüm Ercan, Director of the Şanliurfa Museum, says the scene is a "sky burial," the funeral practice of leaving a body out on a mountain top to be eaten or decomposed which was later practiced in Tibet. The soul, sometimes symbolized as a head, is symbolically carried up to the sky world.

The village of Catalhoyuk is located about 300 miles east of Gobekli Tepe. There were no streets or footpaths. The houses were positioned right up against each other so that people traveled over the rooftops and entered homes through holes in the roofs by climbing down ladders. Although only dwellings have been unearthed so far at the site, geometric designs as well as representations of animals and people are found everywhere. Repeating lozenge shapes and zigzags dance across smooth plaster walls. Hunting parties are shown baiting a wild bull.

Figurines of people were sculpted in clay. The most famous of which is a large woman seated on or between two large wild felines. Other Venus-like figurines often show evidence of having been broken and thrown in the trash, and it is believed that they functioned as wish tokens or to ward off bad spirits. Wild bulls and leopards were also common art subjects.

Golbekli Tepe is not the only place we see home burials and vulture depictions. Deceased relatives were placed under the floors at the Çatalhöyük site. Sometimes the skulls were removed and covered in plaster to resemble live faces. There was also a large mural with vultures and headless humans in one of the buildings, resembling the Golbekli Tepe art.

Ancient cave paintings may or may not have said anything about the afterlife, but vultures carrying way the heads of the dead would certainly indicate that there was some belief in life after death. The treatment of bodies and skulls also suggests that the dead might be seen as influencing the living.

Antequera Dolmens

Around five thousand years after the Anatolian constructions, megalithic dolmens were erected in southern Spain. Dolmens are rock burial structures with a large flat stone laid across the top of upright ones. Modern humans had become experts at burial by then, devoting large amounts of energy and time toward housing the dead. The Spanish Antequera Dolmens property is made up of three megalithic monuments: the Menga Dolmen, the Viera Dolmen and the Tholos of

El Romeral, as well as two natural monuments, La Peña de los Enamorados and El Torcal de Antequera. The three prehistoric burial chambers represent some of the largest and most complete megalithic structures in Europe.

The dolmens were built by farmers who lived in the Guadalhorce Valley. The closest settlement was Cerro de Marimacho to the east, where the large stone blocks were excavated. The blocks were used to create chambers and spaces with linteled roofs or false cupolas to be used for rituals and funerary purposes. The structures were buried in the soil to look like natural caves, and their orientation is based on two natural monuments: La Peña de los Enamorados and El Torcal.

Menga Dolman is the largest dolman in Europe at nearly 30m long. An open atrium section that would have been partially or completely unroofed leads to a passage or corridor section of four stones. This passage leads to the large oval-shaped funerary chamber. When the grave was opened in the 19th century, archaeologists found the skeletons of several hundred people inside. Unfortunately, the paintings at Menga, which were poorly preserved, were erased almost completely when the site was last cleaned.

Viera Dolmen was made of 16 upright stones on each side. Five roof slabs are still in place; originally there would have been three or four more. At the back is a small quadrangular chamber (1.6m wide and 2m high) which is accessed through a small opening cut into the stone. The passage faces east, following the standard pattern of Iberian megaliths.

The Tholos of El Romeral is a false cupola tomb. It has a corridor with masonry walls and a flat roof, of which 11 large stone slabs remain. At the end of the corridor are two initial vertical stones followed by two closer together, to create the impression of a funnel leading into the

5.2m diameter chamber at the end. During the afternoon around the winter solstice the sun shines directly on this back wall. There may have been a dual purpose for the chamber, both for death and astronomical measurements. Maybe the two were linked in some way. We may never know the answer.

Like earlier constructions, these tombs appear to recreate the experience of being in a cave. They are reminiscent of cave paintings made by Neanderthals and ancient *H. sapiens* in European caves. There may be a connection between the two that has been lost in the past. If the dolmens are a reflection of what came before, they may indicate the importance of burials much earlier in human history. And maybe a belief in the afterlife.

Stonehenge and New Grange

Two thousand years after the Spanish megaliths were constructed, people created Stonehenge in England and New Grange in Ireland. Nine miles north of Salisbury, England, the famous prehistoric stone circle named Stonehenge stands on the grassland of Salisbury Plain. DNA tests determined that the people who built Stonehenge were farmers who came from the eastern Mediterranean and hunter-gathers from western Europe. These Aegean farmers lived in Iberia before heading north to Britain about six thousand years ago. A circular earth bank and ditch were the earliest phase of Stonehenge and date to about five thousand years ago. The first large bluestones were erected sometime between four and five thousand years ago.

Scientists believe Stonehenge was both an astronomical observatory and religious site. Gerald Hawkins described in detail how the site could be used to observe the sun and moon over a recurring 56-year cycle. New hypotheses even suggest that Stonehenge could have been a place of healing. There are a high number of burials in the area, with evidence of trauma and deformity in some of the graves. Some of the buried individuals were also from other regions such as the Mediterranean and Germany.

Located north of Dublin, Ireland, the fifty-two-hundred-year-old New Grange passage tomb is the largest of a complex of monuments built along a bend of the River Boyne known as Brú na Bóinne. Passage tombs consist of a passage leading to a chamber where the remains of the dead (usually cremated) were placed. Throughout the region there are as many as 35 smaller passage-tombs and many other archaeological sites built during the Neolithic or New Stone Age by a wealthy farming community.

New Grange consists of a circular mound of stones, 279 feet (85 m) in diameter and 43 feet (13 m) high. Later, it was surrounded by a ring of 35 or more standing stones, of which 12 are still in place. Because it was disturbed before proper excavation, it is not known how many people were originally interred in the tomb. The remains of five bodies were recovered, though the original number was probably much higher. Most of the bones had been cremated, with only small amounts left unburned. Beads made of bone along with pendants and polished stone balls were also found.

Artwork is also abundant. Motifs include spirals, lozenges, chevrons, triangles, and arrangements of parallel lines or arcs. Knowth, another site nearby, has a greater collection of art, with inscriptions on more than 300 of the stones. At both sites it was found that many stones were carved on their undersides and on the sides that are turned inwards toward the cairns. Some suggest that the art could be maps of the area, maps of the stars, or maps of the afterworld. Others see it as a meditation device, maybe enhanced by hallucinogens. The most popular belief is that the forms represent the changing seasons by marking the solstices and the equinoxes. Between December 19 and 23, for a few minutes each morning the sun shines in along the passage and into the burial chamber. Whether this was connected with belief in an afterlife is unknown. It may have housed the spirits of ancestors, providing a link to the spirit world and serving as a focal point for rituals and celebrations.

Dolmens of Marayoor

Along with the large megaliths, dolmens of all shapes and sizes, cists (pitted chambers), menhirs (pointed marker stones), cairns (rock piles), and sarcophagi (stone coffins) have been found around the world.

The Marayoor dolmens of Kerala, India, built about 5000 years ago, were some of the first tombs built in South India. Blocks of rough-hewn granite were organized into small chambers dotting the hilltop. Four plates of granite made the walls of the chamber where the dead, wrapped in leafy shrouds, were situated in a sitting position. A final stone served as the roof. Some dolmens also have an opening in one of the walls to allow the soul to flow back and forth between the body and the aether. Nobody knows exactly who created the Marayoor dolmens. They belong to at least one of many pre-Dravidian tribes—the Malpatarams, the Paniyans, the Uralis, the Kadars, the Muthuvans, or others. Centuries later, The Manimekalai, one of the five epics of Tamil literature, mentions burials such as stone chambers, pits, urns and sarcophagi. Today the dolmens are known as muniyara, or rooms where munis (sages) meditate. Many believe they exemplify their architects' belief in reincarnation.

SECTION IV
THE MYSTERY OF CONSCIOUSNESS

Chapter 14

Consciousness: The Philosophers

If we want to define heaven as the existence of consciousness after the death of the body, it would be helpful to understand consciousness. Galileo ignored the qualities of the physical world in order to give purely quantitative features to matter–size, shape, location, motion–that can be explained with mathematics. But he also suggested that some things exist outside the world of physical measurement. For example, the colors of objects, the tastes in foods, and the smells of flowers all have physical properties. The wavelength of red can be measured, but the quality of red cannot. You can't capture the redness of a red experience or the sourness of lemons on the tongue in an equation. Galileo said that these qualities are "in the soul." The redness of a tomato isn't just the wavelength; it's in the soul of the person perceiving the tomato.

The Conscious Brain

Consciousness is defined as being awake or aware of one's surroundings. It seems like a simple statement, but it is not a simple concept to document. Most believe that not all matter in the universe is conscious. Consciousness is instead restricted to a subset of animals with relatively complex brains. The question is, how complex does the brain need to be? Are only humans conscious? Are other animals just automatons, living their daily lives like a computer running a program?

Ancient societies originally thought the heart was the location of at least part of our consciousness. Greek philosopher, Aristotle, said the heart was the most important organ of the body. It was the seat of intelligence, motion, and sensation. Andreas de Laguna in 1535 wrote, “If indeed from the heart alone rises anger or passion, fear, terror, and sadness; if from it alone spring shame, delight, and joy, why should I say more?" Master Nicolaus, who lived during the late twelfth century, said the heart was the primary "spiritual member" of the body and the seat of all emotions. It wasn’t until the seventeenth century, that French philosopher, Rene Descartes, argued that the heart was a pump and not the seat of emotional thought.

Today we know that the brain, and not the heart, is the site of both thought and emotion. Without going into a lot of anatomical detail, the brain is basically composed of three parts, the cerebrum, cerebellum and brain stem. The cerebrum is divided into four lobes: frontal, temporal, parietal and occipital. Each lobe is involved with different functions like speech, moving parts of the body, seeing, and understanding language. The cerebrum’s outermost layer of nerve tissue is called the cortex or gray matter. It has many folds and grooves and is the location of memory, thinking, learning, reasoning, and emotion. But where is consciousness? Is it spread throughout the brain? Is it a collection of cells in one location?

Following Galileo’s lead, the goal of the philosophy of consciousness was to document the exact relationship between mental states which are subjective and brain states which are physical. A century ago, before the creation of MRIs and other brain scanning equipment, camps were divided into materialists, those who believed that the mind was a product of the physical brain, and dualists, who believed the mind and body were separate. Let's review some of the people who have thought a lot about consciousness.

Bertrand Arthur William Russell

Bertrand Russell was born in Trellech, Monmouthshire, England in 1872 to a family of the British aristocracy. His father, Lord Amberley, was a deist and his godfather was the philosopher John Stuart Mill. By the age of 26, Russell began writing *German Social Democracy* while he taught the same topic at the London School of Economics. A year later he began to study mathematics and wrote *An Essay on the Foundations of Geometry* which focused on non-Euclidean geometry. That was followed by The Principles of Mathematics in 1903 and a co-written, three-volume work with Alfred North Whitehead, *Principia Mathematica* in 1910-1913.

At the age of 29, an event in February 1901 led to what Russell called a "mystic illumination." As he witnessed a colleague's wife suffer an angina attack, he said, "I found myself filled with semi-mystical feelings about beauty...." He continued, "At the end of those five minutes, I had become a completely different person."

Inspired by his student, Ludwig Wittgenstein who published *Tractatus Logico-Philosophicus* in 1922, Russell became known for being one of the founders of analytic philosophy. He wrote about many fields, including metaphysics, logic and philosophy of mathematics and language, ethics, and epistemology. His monist theory of metaphysics stated that a single set of properties underlies both consciousness and basic particle physics. Russell argued that there must be more to physical entities than what they do. He suggested that physical things must also have an intrinsic nature.

Since physics is restricted to only measuring the interactions of physical particles, it has no way to measure intrinsic nature. He wrote, "All that physics gives us is certain equations giving abstract properties of their changes. But as to what it is that changes, and what it changes from and to—as to this, physics is silent."

In his book, *The Analysis of Matter*, he outlined his novel approach to the consciousness problem. His view contained two components. First, the physical sciences are limited to describing the external nature of matter in relational and mathematical ways. Second, science cannot explain the intrinsic, concrete or categorical nature of things, especially consciousness. Some Russellian monists believe that the intrinsic nature of matter is "consciousness-involving." Other believe that it involves unseen properties that somehow explain the reality of consciousness.

Although Russell proposed this unique approach to the question of consciousness, Russellian Monism was almost completely forgotten for much of the twentieth century.

Arthur Stanley Eddington

Arthur Eddington was born in 1882 in what is now called Cumbria in England to Quaker parents, Arthur and Sarah Eddington. His father was headmaster of Stramongate School, but he died of typhoid when Arthur was only two years old, leaving his mother to raise two children alone. Arthur was educated at home before attending three years at a preparatory school. In 1893, he entered Brynmelyn School, where he excelled in mathematics and English literature. He earned a scholarship to Owens College, Manchester at the age of 16. After spending the first year taking general courses, he focused on physics for the next three years. He graduated with a BSc in Physics with First Class Honors in 1902.

After attending Owens College, he was awarded a scholarship to Trinity College, Cambridge. He received his M.A. in 1905 and began research on thermionic emission (charged particles emitted from an electrode), later securing a position at the Royal Observatory, Greenwich, which led to a career in Astronomy. By 1906, he was nominated for the post of chief assistant. He pursued important studies in astrophysics and relativity as well as teaching and lecturing. His greatest contributions in the field of astrophysics included pioneer work on stellar structure and radiation pressure, subatomic sources of stellar energy, stellar diameters, the dynamics of pulsating stars, the relation between stellar mass and luminosity, white dwarf stars, diffuse matter in interstellar space, and spectral lines. In 1919 he led an expedition to Príncipe Island (West Africa) and provided the first confirmation of Einstein's theory that gravity will bend the path of light when it passes near a massive star.

Eddington believed that quantum theory and general relativity could be united. He felt that it was possible to calculate the values of universal constants. He attempted it, but it was never complete and was published posthumously as *Fundamental Theory* in 1946. Eddington also proposed philosophical ideas that together with Russell brought Panpsychism into the 20th century.

Panpsychism

Panpsychism is an ancient concept that predates Greek philosophy and has resurfaced throughout history. It is the view that all physical things have a mind-like quality. The word was coined by Italian philosopher Francesco Patrizi in the sixteenth century and derives from two Greek words *pan* (all) and *psyche* (soul or mind). Sounds like Animism doesn't it?

Unlike Galileo who chose to ignore the mind and focus only on the physical, Panpsychism is the view that consciousness is a fundamental part of the natural world like gravity or electromagnetic fields. Modern physics offers a simple and unified vision of the physical world, but it has yet to account for the emergence of consciousness. Panpsychism is not a final hypothesis of consciousness but may be a framework for making theoretical progress. Modern philosophers, like Philip Goff, are trying to unite the mind and body by looking at consciousness as an inherent physical force.

Philip Goff

Philip Goff was raised as a Catholic and admits in interviews to being bothered by the mind-body dualism taught by the church. When he pursued a degree in philosophy as an undergraduate at Leeds, he felt he had to choose between dualism (the belief that the mind and brain are separate) and materialism (the mind is dependent upon the existence of physical brain). He became a committed materialist and debated many religious leaders, insisting that the mind and the brain were the same thing. There was no spirit or soul.

Over time, Goff's outlook on the universe changed. "I couldn't deny it anymore," he said. "I'd already accepted that if materialism was true, then I was a zombie. But I knew I wasn't a zombie; I was a thinking, feeling human being. I could no longer live in denial of my consciousness." Goff finished his studies as a "closet dualist," arguing in his dissertation that defining consciousness was unresolvable. Dissatisfied with his entire field of study, he went on to teach English in Poland.

It wasn't until later that he discovered an article on Panpsychism written by Thomas Nagel. Perhaps there was a third way to think about consciousness. Goff obtained his PhD at the University of Reading after studying the writings of both Russell and Eddington and others. His postdoctoral work took place at the Centre for Consciousness at the Australian National University, and he is currently an Associate Professor at Durham University.

Goff's third view of consciousness, a version of Russellian monism, attempts to account for consciousness' intrinsic nature by suggesting it is a fundamental feature of the physical world. "The basic commitment is that the fundamental constituents of reality—perhaps electrons and quarks—have incredibly simple forms of experience." Because the claim is that consciousness is fundamental, Goff believes that consciousness cannot be reduced to something else.

In his book *Galileo's Error*, Goff uses several arguments to show that a purely physical description of the world is not complete, and that materialism cannot be entirely true. Some believe he is trying to say that electrons can think. Goff explains that most modern Panpsychists do not believe that clumps of matter result in complex consciousness. He says, "Einstein tells us weird things about the nature of time that counters common sense; quantum mechanics runs counter to common sense. Our intuitive reaction isn't necessarily a good guide to the nature of reality."

This might sound like a bunch of mumbo-jumbo for a book that talks about the importance of hypothesis testing. Can we test for the presence of consciousness in the universe? Not at this point in time. Is it possible that we might be able to find the location of consciousness in the brain? Maybe. Traditional explanations of consciousness continue to fail, and the Panpsychist view is being taken more seriously by credible philosophers, neuroscientists, and physicists, including figures such as neuroscientists Christof Koch and Frances Crick, Morton Overgaard, philosopher David John Chalmers and physicist Roger Penrose.

Chapter 15

Consciousness: The Neuroscientists

The Self-Aware Brain

Using brain-imaging technology, neuroscientists first discovered that the cerebral cortex is critical for self-awareness, particularly three areas: the medial prefrontal cortex, anterior cingulate cortex, and insular cortex regions. The medial prefrontal cortex is located at the front of the brain. It is connected to deeper regions in the brain that are associated with emotional processing (amygdala), memory (hippocampus) and sensory input (within the temporal cortex). The anterior cingulate cortex connects with both the emotional (limbic) system and the prefrontal cortex. The insular cortex plays a role in diverse functions linked emotion and the regulation of the body's homeostasis, such as hunger, pain and fatigue. But that is not the whole story.

New evidence collected by a University of Iowa research team showed that self-awareness is more a product of a patchwork of pathways in the brain rather than being confined to three specific areas. Their conclusions came from a rare opportunity to study a person with extensive brain damage to the areas mentioned above. The person, a 57-year-old, college-educated man, passed all standard tests of self-awareness. He recognized himself when looking in the mirror and at photographs taken during all periods of his life. They concluded that self-awareness may just emerge from more diffused interactions in areas of the brain, including the brainstem, thalamus, and posteromedial cortices. David Rudrauf, the co-author of the paper, states, "What this research shows is that self-awareness corresponds to a brain process that cannot be localized to a single region of the brain."

In a 2019 study, researchers collected fMRI data from 159 people and found that the brains of healthy individuals had more complex and changing patterns of signaling than unconscious people. They also found that dopamine improves self-awareness. This may be why relaxation meditation like yoga or mindfulness meditation, which have shown to increase dopamine, also improve awareness.

At this point in time, we are just beginning to understand the workings of the physical brain. Is there a way to test if consciousness is produced by the brain or if it is something separate from the physical neural tissue?

Neural Correlates of Consciousness

Scientists have been trying to find the neural correlates of consciousness (NCC), which is defined as the minimal amount of brain matter needed to have a conscious experience. They have been able to eliminate parts of the brain as locations of consciousness. The cerebellum oversees coordinating motor movements and motion. If it is injured or damaged by a stroke or surgery, there is no loss of conscious thought. The patients feel fine, retain their sense of self, remember past events, and plan for the future. As important as the cerebellum is, it is not the seat of consciousness.

Surgery on the frontal lobe of patients who had epileptic seizures or tumors removed has also given us clues to the location of conscious awareness. The loss of some of the frontal lobe could result in motor control problems, lack of inhibition, inappropriate emotions or repetition of specific action or words. However, the patient's personality and IQ usually improved, and there was no effect on consciousness.

Before conducting brain surgery today, neurosurgeons map the functions of nearby tissue by stimulating it with electrodes. They have found that stimulating the front of the cerebral cortex elicits no experiences. Conversely, stimulating areas in the posterior part of the cortex can trigger a diversity of sensations, including flashes of light, geometric shapes, auditory or visual hallucinations, urges to move limbs, and feelings of familiarity or unreality. Removing even a small portion of the posterior area can lead to a loss of entire groups of conscious content like motion, color, space or facial recognition. At this point in time, scientists don't know why there is a crucial difference between the posterior and prefrontal cortex. Are they closing in on the physical location of conscious thought? Will they be able to measure it in the future?

Neurologist, Christof Koch wrote:

> *Why these neurons and not those? Why this particular frequency and not that? Indeed, the abiding mystery is how and why any highly organized piece of active matter gives rise to conscious sensation. After all, the brain is like any other organ, subject to the same physical laws as the heart or the liver. What makes it different? What is it about the biophysics of a chunk of highly excitable brain matter that turns gray goo into the glorious surround sound and Technicolor that is the fabric of everyday experience?*

Many scientists and philosophers agree that at present we don't have a way to explain how conscious experience is generated from the firings of neurons. David John Chalmers, an Australian philosopher specializing in the mind and language, called it "the hard problem of consciousness." The physical brain contains neurons that communicate with one another. Research has found areas for example that can identify faces, but how do we get from "area X equals faces" to "I think, therefore I am?" Frances Crick and Christof Koch tried to answer that question.

Francis Crick and Christof Koch

Frances Crick was born in 1916 near Northampton, England to Harry Crick and Annie Elizabeth Crick. His father and uncle ran a boot and shoe factory, while his grandfather was a naturalist who corresponded with Charles Darwin. Crick followed in his grandfather's footsteps. At an early age, he was attracted to science. By the age of 12, he refused to attend church anymore and preferred a scientific search for answers over religious belief.

Crick attended University College, London and earned a BS in physics from by the University of London in 1937. Although his PhD studies were interrupted by World War II, he became a PhD student and Honorary Fellow at Gonville and Caius College, and later worked on molecular biology in Cambridge. He became a research student for the second time in 1950 at Caius College, Cambridge, and obtained a PhD with a thesis entitled: *X-ray diffraction: polypeptides and proteins.* After befriending J. D. Watson in the early 1950s, the two, working with Rosalind Franklin, identified and proposed the double-helical structure for DNA. They then suggested a general theory for the structure of small viruses.

Toward the end of his career, he held the post of J.W. Kieckhefer Distinguished Research Professor at the Salk Institute for Biological

Studies. He focused his research on theoretical neurobiology and attempted to advance the study of human consciousness with Christof Koch.

Christof Koch was born in Kansas City, but grew up in Hollard, Germany, Canada, and Morocco where he received his baccalaureate in classical French from the Lycée Descartes in Rabat in 1974. He studied physics and philosophy in Germany, receiving a master's in physics from the University of Tübingen, Germany and his Ph.D. from the Max-Planck Institute for Biological Cybernetics in 1982. The next four years he spent as a postdoctoral fellow in the Artificial Intelligence Laboratory and the Brain and Cognitive Sciences Department at the Massachusetts Institute of Technology. From 1987 -2013, Koch was a professor at the California Institute of Technology (Caltech) in Pasadena, from his initial appointment as Assistant Professor, Division of Biology and Division of Engineering and Applied Sciences in 1986, to his final position as Lois and Victor Troendle Professor of Cognitive & Behavioral Biology.

During part of his tenure, Koch was a member of a team of 300 scientists at the Allen Institute for Brain Science. One project had the goal of identifying and cataloguing all the types of neurons in human and mouse brains. The second project was to create a sort of brain observatory by using optical and electrical recording devices to record brain-wide cellular activity. Koch's interest was in perception, memory, behavior and consciousness.

Collaborating with Francis Crick, he began a modern search for Neural Correlates of Consciousness (NCC). Their hypothesis was that the nerves of the brain give rise to all conscious experience, and because of that scientists should be able to measure the connection. For example, when a person sees a face, what neurons fire? If those neurons are stimulated with TMS transcranial magnetic stimulation, does the person also see a face?

The problem with finding NCC is that brains contain about 100 billion neurons. And to make things more complicated, brain cells have projections called dendrites and axons that connect cells to each other in ways unique to every individual. There is also redundancy and parallelism in networks of nerves, making the brain anatomy too complex to test for cause and effect. It wasn't easy to determine what characterizes NCC. They had many questions. What are the commonalities between the NCC for seeing and for hearing? Do the NCC involve all the pyramidal neurons in the cortex at any given point in time or only a subset. Do the neurons fire in a rhythmic or synchronous manner?

In a 2016 review, Koch noted that measuring certain brainwave activity (Gamma and P3b waves) has no predictive power. "In short, no single brain area seems to be necessary for being conscious, but a few areas, especially in the posterior cortical hot zone, are good candidates for both full and content-specific NCC."

Morten Overgaard

Morten Overgaard received a master's in psychology from the University of Copenhagen and a PhD from Aarhus University. He founded the Cognitive Neuroscience Research Unit and became a Professor in Cognitive Neuroscience. Since then, he has led many large-scale European research projects. He is a pioneer in translational neuroscience, investigating how subjective consciousness relates to behavior and brain processes. His main goal is to help people improve their cognitive functioning after a brain injury.

Overgaard wrote that scientists generally agree that consciousness is subjective. Consciousness research is able to provide meaningful and reproducible results but has yet to explain why and how consciousness exists. When it comes to measuring consciousness, direct approaches simply report a person's experiences, but because this is subjective, it may not be reliable. Scientists don't like to use personal reports, so some try indirect approaches where the subject chooses between two options which were established by the original subjective reports. But that option is not entirely reliable.

There has been an increase in the number of studies in the area, but the fact remains that consciousness is subjective and only accessible from the first person. We have no accepted measures of consciousness, and the main challenge is to measure it from the outside. According to Overgaard, "consciousness remains one of the biggest scientific challenges among all disciplines as the most fundamental questions are not simply unanswered–it is still highly unclear how one should even begin to answer them."

Integrated Information Theory

Physicians working with unconscious or comatose patients are often at a loss when it comes to judging consciousness. Giulio Tononi was born in Trento, Italy, and obtained an MD in psychiatry and a PhD in neurobiology at the Sant'Anna School of Advanced Studies. He holds the David P. White Chair in Consciousness Science at the University of Wisconsin. As an authority on the genetics and etiology of sleep, he has worked with fruit flies and rodents to develop models.

Tononi and Marcello Massimini in 2013 pioneered a technique called transcranial magnetic stimulation (TMS) to probe whether a patient is conscious or not. They induced a magnetic pulse into the skull, creating a brief electric current in the brain cells beneath. The cells responded by sending signals across the brain while the scientists recorded the activity with electroencephalogram (EEG) sensors. In the conscious brain, the signals were complex and spread widely over the surface of the cerebral cortex. If a patient was under general anesthesia, or in a dreamless sleep, the signals produced were simple. After being standardized, this technique could be a tool that can detect consciousness even in people who can't see, hear or respond to commands.

Tononi developed the Integrated Information Theory (IIT) which attempts to explain what consciousness is, how it can be measured, and how it is correlated with brain states. It is a mathematical model for the consciousness of a system and is made of a framework to explain why some physical systems (such as human brains) are conscious.

Instead of focusing on brain neurons, IIT takes the opposite approach. It begins with the experience, by identifying its essential properties, and then infers what kind of physical systems must be present to account for those essential properties. It then uses the postulates to derive, for any particular system, whether it has consciousness, how much, and of which kind. From these premises, IIT makes testable predictions.

Like any hypothesis, IIT needs to be tested. Tononi is testing his model with neuroimaging, transcranial magnetic stimulation (TMS), and computer models. "According to IIT, a system's consciousness (what it is like subjectively) is conjectured to be identical to its causal properties (what it is like objectively). Therefore, it should be possible to account for the conscious experience of a physical system by unfolding its complete causal powers." So far, this system looks at consciousness from a different point of view, but does not really solve the problem of defining consciousness?

Meditation and the Brain

Not only are humans conscious and self-aware, but we have also discovered that the brain can control its own conscious state. Meditation has been around for centuries, but only recently have scientists documented its effects on the brain. The practice can change not only the brain's functioning but the physical structure of the prefrontal cortex, the amygdala which controls emotion, and the hippocampus which is responsible for learning and memory.

Most people have heard that meditation can reduce stress. Scientists have found that study participants trained in either mindfulness meditation or a relaxation training program show improvement in a blood marker responsible for inflammation, which is linked to stress. Another study found that meditation can also improve attention and concentration and decrease the mind's wandering.

Scientists at Harvard tested people who had never meditated before to see if the practice of Mindfulness-Based Stress Reduction (MBSR), had any effect on the physical brain. First, they collected MRI images from all the participants, and then half of them began an eight-week meditation program. The other half were control subjects who did no meditation. The results showed that those who meditated had changes in gray matter concentration in brain regions involved in learning, memory, emotion regulation, and self-awareness. The participants who showed brain changes also reported they felt less stressed as well.

There is also research that suggests that meditation might protect your brain from aging. In one study, researchers compared the brain scans of 50 people who meditated regularly for 20 years with those who didn't. Both groups showed a loss of gray matter, but there was less of a decline for those who meditated.

If meditation can change the physical make-up of the brain, what about people who meditate all the time? In 2008, New York University neurology researchers imaged the brains of over 20 Buddhist monks during meditation with MRI. Dr. Zoran Josipovic attempted to understand how their brains reorganize during the process. He found that meditation can beneficially change the inner workings and circuitry of the brain.

Richard Davidson, at the University of Wisconsin School of Medicine and Public Health and the Waisman Center also worked with monks. In collaboration with colleagues Julie Brefczynski-Lewis and Antoine Lutz of the UW-Madison W.M. Keck Laboratory for Functional Brain Imaging and Behavior, Davidson compared newly trained meditators to people with up to 54,000 hours of meditation experience.

After the novices were taught to meditate, all subjects underwent an MRI scan of the brain while they were meditating. The scientists found greater activity in brain connections involved with paying attention in the monks and newly trained meditators. They found that regions of the brain that are involved in the control of attention, such as the prefrontal cortex, were more activated in the monks, and that the monks were able to concentrate in an effortless way.

Overall, meditation appears to have several effects on the brain. The happier parts of the brain (prefrontal cortex) are far more active. The brain tends to reorganize itself in such a way that it feels a sense of "oneness" with the world. The brainwave patterns of the Buddhist monks suggested a higher level of enhanced focus, memory, learning, consciousness, and neural coordination. This may account for the monks' lack of anxiety, depression, and addiction.

Think about what the research is finding. Humans are not only conscious and self-conscious, but they can also physically change the inner workings of their brains and change connections between neurons just by thinking about it. The question is: does conscious thought, that can alter the brain, extend beyond the brain as a physical organ? One can only know for sure by testing the hypothesis.

The Wager

A headline in *Nature News,* published 24 June 2023 announced, "Decades-long bet on consciousness ends-and it's philosopher 1, neuroscientist 0."

Christof Koch (neuroscientist) and David Chalmers (philosopher) made a bet in 1998 that the way the brain's neurons produced consciousness would be discovered. They made the bet at the annual meeting of the Association for the Scientific Study of Consciousness in New York City.

Both were involved with a large project supported by the Templeton World Charity Foundation. Their goal was to set up a series of opposing experiments to test various hypotheses about consciousness. The findings from one experiment tested two leading hypotheses: IIT and global network workspace theory (GNWT). IIT says that consciousness is a structure in the brain formed by a specific type of nerve cell connectivity. It is believed to be in the posterior cortex at the back of the brain. The GNWT says that consciousness is formed when information is broadcast to areas of the brain through an interconnected network. This involved the prefrontal cortex, at the front of the brain.

The results, which weren't yet peer reviewed at the time of the 2023 meeting, didn't match either of the hypotheses. The IIT experiment showed that the posterior cortex does contain information in a sustained way, but there Is no synchronization between other areas of the brain. For the GNWT, found some aspects of consciousness in the prefrontal cortex, but the broadcasting part was only at the beginning and not end as predicted.

Koch thinks that in another 25 years they will have the answers.

SECTION V
EARLY ACCOUNTS

Chapter 16
Near-Death Experiences

Did the belief in heaven begin with visions? Was it created by some ancient person's imagination? Maybe it was a hallucination experienced by a shaman. Near-death Experiences (NDEs) became a popular topic in the 1970s when Dr. Raymond Moody published a book on the subject. Modern medicine and the ability to resuscitate patients after their hearts stopped, certainly led to an increased number of people remembering NDEs. These experiences included the patient leaving their body and rising upward. They would pass through a dark enclosure, sometimes described as a tunnel until they reach a brilliant white light. The bright light brings them comfort, joy, peace and love. They also may see deceased loved ones who are there to welcome them and they experience events of their lives in a full-color panoramic way.

We think of the NDE as a modern experience, but that is not the case. Even without modern medicine, ancient people were having both visions and experiences of traveling to the afterworld.

India

The *Rig Veda,* dating to at least thirty-five hundred years ago, contains an account of a boy who traveled to the realm of the dead in a "chariot of the mind" to visit his recently deceased father, but this is a vision, not a near death experience. In another story found within the *Shatapatha* and *Jaiminiya Brahmanas,* written about twenty-eight hundred years ago, the son of the deity Varuna is sent to the otherworld to gain knowledge, and returns.

The story of Naciketas, which was recounted in the *Katha Upanishad* about twenty-five hundred years ago, may be the first account of a near death experience ever recorded. The story begins with Naciketas's father becoming so angry at his son that he "gives him to death" or kills him. Naciketas waits three nights for Yama, the god of death, to arrive. Because Yama is late, he gives Naciketas three wishes. Naciketas asks to return to life and that his father no longer be angry with him. He also asks Yama to clarify whether a person exists in the afterlife or not. Yama explains the importance of good behavior, the principle of rebirth, and the identification of the inner unchanging self (*atman*) with the transcendent or ultimate reality (*brahman*) before he returns Naciketas back to his body where he eventually becomes a religious teacher.

Greece

Plato recounts the story of a soldier, Er, who died about twenty-four hundred years ago. Er spontaneously revived on his funeral pyre and recounted his near-death experience. According to the account, while Er was thought to be dead he descended to the underworld. Along with other spirits, he came across a strange place with four openings. Two led into and out of the sky and two into and out of the ground. Judges sat between these openings, directing individuals which path to follow. The good were guided along the path into the sky; the bad were directed below. When Er reached the judges, he was told to remain where he was, for he was to report his experience to humankind.

Er witnessed "clean" souls floating down through the sky openings, recounting wonderous sights and feelings. Souls returning from underground were dirty and haggard, describing their awful experiences and that they had to pay tenfold for their misdeeds. Murderers, tyrants, and other criminals were doomed to remain in the underground for eternity.

Over the next several days, Er described the afterworld in detail, including a lottery where each soul was required to come forward to choose his or her next life. Many preferred a life different from their previous experience. Animals were present who chose human lives, while humans often chose the apparently easier lives of animals.

After the lottery, each soul was assigned a guardian spirit, and they passed to the River of Forgetfulness where they were required to drink some of the water and not remember anything that came before. Each soul was lifted into the night to complete their journey. Er remembered nothing of the journey back to his body, but he vividly recalled his journey to the afterlife.

In another account, Plutarch wrote the story of Thespesius of Soli who died of a blow to the head but revived during his funeral three days later. Thespesius was a young man leading a corrupt life who had become wealthy through his unjust dealings with others. When he went to the oracle of Amphilochus and asked the deity whether he should try to live a better life, the oracle said things would be better when he was dead. Not long after his visit, he fell from a precipice, landing on his neck. There were no wounds or broken limbs, but he stopped breathing. After three days he was carried to his grave. Just as they were lowering his body, he returned to life.

He described his journey to the afterworld. At first, he felt as if he had been thrown into a storm amid the sea. He slowly rose through the water to find himself among the stars, each "sending forth a light most wonderful for the brightness of its color." He said his soul traveled along this light as if he were in a chariot. He could see souls rising from below him like fiery bubbles which burst, turning into people who seemed troubled. He tried to speak to them, but they didn't hear him. Other souls towered above him. These looked happy and kind and shunned the others below him.

After some time, he recognized a person he had known when he was very young. The spirit guide said that Thespesius was not dead yet and had only come there with his intellectual faculty. His soul was still anchored to his body. The man spoke in detail of the three punishments given out by three of several Furies (female ministers of justice). Thespesius was shown various colors of souls. Black represented avarice and fraud. Bloody and flame-like colors represented cruelty and desire of revenge. Blue was a sign that a soul was barely cleansed from the impurities of pleasure and voluptuousness. Dark violet represented envy.

Thespesius was taken to various places. He travelled along the light to a gaping chasm. Other souls were there, flying around the gaping rift but unable to escape it. Next, he was transported to a large opening where he saw a goblet, into which several rivers flowed. One was whiter than snow. Another resembled the purple color of the rainbow. The spirit guide tried to show him the light of the Tripod, but Thespesius could not see anything because of the extraordinary brightness. He could only hear the shrill voice of a woman speaking in verse to foretell the time of his death.

Thespesius spent much of his time witnessing the torments of those who were being punished. He saw acquaintances and companions groaning under horrific pains. They called him by his name. Then he saw goblins and demons torturing souls in lakes of molten metal. The last thing that he witnessed were souls being transformed into all sorts of creatures by the force of tools and anvils for a second life.

Toward the end of the experience, a woman grabbed his arm, said to him: "Come hither, that thou mayst the better be enabled to retain the remembrance of what thou hast seen." He whirled away in a strong, violent wind and was forced through a pipe back into his body.

Middle Ages

In the 6th century, St. Gregory the Great included three near-death experience tales in the fourth book of his *Dialogues*. In the first story, a hermit who awakened from death reported that he visited hell. While there he saw powerful men dangling in fire. Before he was dragged in into the fire himself, an angel rescued him and warned him to "consider carefully how you will live from now on." The hermit was transformed by the experience.

The second account tells of a businessman named Stephen who does not believe in hell. When he temporarily dies, he gets to see it for himself and changes his view of heaven and hell. Ironically, it appears that this is a case of mistaken identity. The person who should have gone to hell was Stephen the blacksmith. They had died at the same hour.

In the third narrative, a Roman soldier was struck by plague. He spoke of seeing a bridge over a black, gloomy river with a foul smell. If a person was unjust, he would slip and fall into the water, but if he was just, he could cross freely. Across the bridge there were people clothed in white meeting in meadows carpeted with green grass and sweet-smelling flowers.

Another account comes from St. Bede the Venerable who was born about 672 in Northumbria, England. He was an Anglo-Saxon theologian, historian, and chronologist. He is best known for his *Ecclesiastical History of the English People* which helped convert the Anglo-Saxon tribes to Christianity. In the 8th century, he described the near-death experience of a man named Cunningham or Drythelm who "rose from the dead" in 696 A.D.

Drythelm was a devout Northumbrian man who died of an illness. According to the account, the dead Drythelm suddenly recovered, frightening everyone in his household. He scared them further by relating his tale of visiting the afterlife to his wife. First, was a tour of hell, and then a tour of heaven.

A spirit guide of "shining countenance" led him into a vast valley with flames on one side and hail and snow on the other. Countless souls had been tossed into the center. The guide explained that they could be saved if the living undertake pious actions on their behalf.

The guide takes him to hell, a bottomless, stinking pit with tongues of fire lapping at the souls who fall helplessly into it. They next travel to the top of a vast wall and find a flowery meadow amidst a realm of clear light. This is not heaven, but an antechamber. When he reaches heaven, Drythelm is thrilled by the sights, fragrances and beautiful singing. He wants to stay but is sent back to life, transformed. He gives away all his property and retires to a monastery.

In his ecclesiastical history, Bede also included two other deathbed visions. All these accounts stressed living an honorable life to avoid punishment in the afterlife.

Other near-death visionaries include Alberic of Settefrati, a monk born circa 1100. He had a vision at the age of ten during a nine-day illness where he was thought to have died. Guided by St. Peter and two angels, he visited hell, crossed a river of purgatory, and was given a tour of the seven heavens and the land beyond, of which he is forbidden to speak before returning to his body.

Another visionary, Adamnan, is given a tour of heaven by a guardian angel and sees the Glorious One upon a throne enveloped in music and fragrance within the seven-walled heavenly city. There is a second city for those not ready for heaven, and a hell, where souls are punished and devils live beyond a fiery wall.

Sinful Irish knight Tundale's vision was written in 1149 by an Irish monk. While thought to be dead, a guardian angel leads him through sections of hell first. He then proceeds upwards, to fields and pavilions which contain successively purer souls. At the end, he meets St. Patrick and begs the angel to let him stay but is sent back as punishment. He awakens a changed man.

Eighteenth Century Physicians

Chartier wrote in *Resuscitation* about the experience he translated from the French written by Pierre-Jean du Monchaux (1733-1766), a military physician of northern France, who died of fever at the age of 33 in San Domingo. In a description given by the physician, Sir L.C. , about 1740, the patient had a malign fever and was treated by French physicians and surgeons with blood-letting. After the last treatment, he had syncope (drop in heart rate and blood pressure) and was unconscious for such a long time that they were worried he wouldn't recover. When he regained consciousness, he reported having lost all external sensations. He saw a pure extreme light that he thought was heaven. He claimed to never in his life have had such a pleasant moment.

George de Benneville was a physician and minister born in 1703. He traveled to France and eventually settled in Germany, where he had his near-death experience at the age of 36. Afterward, he emigrated to America, arriving and settling in the Philadelphia area. He built a large house that contained a schoolroom for both immigrants and Native American children. It was also used by many religious groups and functioned as a space for his medical practice. He assisted in publishing the first German language edition of the Bible in the United States and advocated the doctrine of Universal Salvation that all creatures would be restored to what he called "happiness and holiness" at the end of time.

In 1741, he wrote a first-person near-death experience account. His account included his description of leaving his body:

> *I felt myself die by degrees, and exactly at midnight I was separated from my body, and saw the people occupied in washing it, according to the custom of the country. I had a great desire to be freed from the sight of my body, and immediately I was drawn up as in a cloud, and beheld great wonders where I passed, impossible to be written or expressed.*

He arrived in a heaven-like location with winged guardians:

> *I quickly came to a place which appeared to my eyes as a level plain, so extensive that my sight was not able to reach its limits, filled with all sorts of delightful fruit trees, agreeable to behold, and which sent forth such fragrant odours that all the air was filled as with incense. In this place I found that I had two guardians, one at my right hand and the other at my left, exceeding beautiful beyond expression, whose boundless friendship and love seemed to penetrate*

> *through all my inward parts. ... They had wings and resembled angels, having shining bodies and white garments.*

Guardians took him on a tour of the "seven habitations of the damned:"

> *You might see fornicators, idolaters, adulterers, thieves, the covetous, drunkards, slanderers, ravishers, etc., each laboring and being employed with his sins and iniquities. One might also see all kind of conditions of men, divines, deputies, controvertors, advocates, judges, lawyers, and in a word one might discover whatsoever any of them had done upon earth. In each habitation I discovered that those who were abased and that appeared sorrowful for their sins, were separated from the others of seven habitations of the damned, where I knew one I had been acquainted with upon earth.*

He was lifted from the place by a messenger:

> *Beholding the messenger attentively, I discovered that he had a most glorious body, dressed in a robe whiter than snow, filled with the most exalted love and friendship, joined with the deepest humility which penetrated me through and through, and suddenly there was heard a great multitude of the heavenly host, and the messenger said, as he flew to join the same, with a sweet voice – "Holy, holy is the Lord God Almighty, who was, and is, and who is to come."*

Then he traveled to a heavenly place with five mansions:

> *The messenger then came and conducted us into a most wonderful place, and ordered my two guardians to conduct me into five celestial mansions, where the Lord's elect abide; and then to reconduct me to dwell yet a time, and half a time in my earthly habitation, and to preach to the lower world the universal everlasting gospel; and that the most holy trinity hath a pure universal love towards all the human race, without exception, and to each one in particular; then turning himself towards me he said, my dear soul, my dear brother, thou shalt be favored of the most holy trinity, to be conducted by thy guardians, who shall never leave thee; when thou shall have need of their counsel, thou needest but to call them, and they shall be day and night present at thy service; they shall conduct thee into five of the heavenly mansions at this time, where thou shalt partake in a certain degree, of the celestial glory as much as thy spirit shall be able to receive, as not being yet sanctified and purified sufficiently, and then thou shall be reconducted into thine earthly tabernacle, for a time and half a time and shall preach to the lower world the universal everlasting gospel, and that the most holy trinity hath a pure universal love towards all the human race without exception, and towards each one in particular.*

He continued to describe his experience in detail, including:

Many thrones, palaces, edifices, temples, and buildings were erected in all parts, with fruit trees intermixed, rivers of pleasure gliding along through the celestial land, which appeared like a garden of heaven, even the paradise of God. It is the court of the King of Kings, and Lord of Lords, which the eye has not seen nor the ear heard, and which the hearts of men have not received. It is the celestial globe where the New Jerusalem, or Mount Sion, is placed, where the bosom of Abraham is; where the sufferers who came out of their tribulations are refreshed, and rejoice forgetting all their miseries; being come out of their purifications, they are made to rejoice in Sion; O magnificent globe! O thou city of the Great God! stately city of this place! where shall a mortal find convenient phrases to lift out a little of thy glory and splendor? It is the glory and magnificence of the most holy trinity, where God is pleased to manifest himself in his pomp and beauty. The blessed angels have their employment in serving God; they compose the court of the Great King. O my God, I am not able to express that which penetrates me, of the grandeur, magnificence, splendor, pomp and majesty of thy dwellings, or of the inhabitants in those transparent places, hallelujah and victory for ever. ... AMEN.

At the end, the spirit guardian returned him to his body:

Then my guardian took me up, and reconducted me to the house from whence I came, where I perceived the people assembled, and discovering my body in the coffin, I was reunited with the same, and found myself lodged within my earthly tabernacle, and coming to myself, I knew my dear brother Marsey, and many others, who gave me an account of my being twenty-five hours in the coffin, and seventeen hours before they put me in the coffin, which altogether made forty-two hours; to me they seemed as many years; beginning then to preach the universal gospel, I was presently put in prison, but soon set at liberty again. I visited all my brethren, preaching the gospel and taking leave of them all, because that my God and Sovereign Good called me to go to America and preach the gospel there. I took my departure for the same in the 38th year of my age, and it is forty-one years since I first arrived here. The 28th of July next, 1782, I shall be 79 years of age. Blessed be the name of the Lord forever.

Chapter 17
Seers

Not only were people experiencing NDEs in times past, but seers were also recording accounts of their visions of heaven. This chapter discusses several well-known seers from the Middle Ages to the 20th Century. These are not all the seers, but they convey thoughts about the afterlife and heaven.

Middle Ages

There were several stories about visions of heaven during the Middle Ages. Most of these stories were written by monks who may have had ulterior motives of keeping their flocks in line. Most of the tales focus on suffering in hell and being a good person.

During the twelfth century, according to legend, there was a small island in the red waters of a lake in northwestern Ireland that held a cave which led to the afterworld. It was said that St. Patrick used it to convert the Irish pagans in the fifth century. By the late Middle Ages, thousands of pilgrims flocked to the cave. They would be laid out inside as if they were dead, given last rites and locked in, in the hope of witnessing the afterworld. The cave was destroyed in the eighteenth century, but some accounts still exist.

One vision comes from an account written by the English Cistercian monk, H of Sawtry. The pilgrimage occurred at the end of the twelfth century and is recorded in the *Treatise on the Purgatory of St Patrick.* It tells us about Sir Owen, a knight of the Crusades.

Owen was remorseful for having devoted his life to violence and plunder and traveled to the cave as penance. During his visit, he had a vision. It began with Owen traveling along a dark passageway. At the end, he met twelve white-clothed men who instructed him to use the name of Christ as a protective prayer for the rest of his journey. After they vanished, he witnessed several disturbing events. He first heard multiple creatures screaming. Then he saw sinners eaten by dragons, attacked by snakes and toads, and cooked in furnaces and cauldrons. When he reached a well of flames, he witnessed naked souls shooting out like sparks and was told this is the mouth of hell. Then they threw him in. He fell endlessly, forgetting at first to call on Christ. When he finally remembers to call for help, he is lifted to safety.

Owen then traveled to a river of fire spanned by a narrow, slippery bridge. As he crossed the bridge, it grew wider, and his sins were cleansed. He passed through a jeweled gate into a land of dazzling light, where he met a procession of holy men singing in harmony. He toured beautiful meadows whose flowers release fragrances, on which, he felt, he could live forever. He discovered this is only the earthly paradise where purified souls await God's call to heaven. Two archbishops guided Owen to the top of a mountain. A flame descended from a golden sky, entering and nourishing each person with "heavenly sweetness."

Owen wanted to stay forever but was told he must go back. It would be possible to return if he lived a good life after this event. Once locked out of the jeweled gate, he retraced his steps and reached the exit door at dawn, just in time to be released. Owen was transformed. He went on a pilgrimage to Jerusalem and took up the religious life, serving the Cistercian order.

Another vision was written in 1197 by the Monk of Evesham, brother of Edmund, the Benedictine monk who reported the vision. His fellow monks found him and thought he was dead on Good Friday, but when he revived at midnight before Easter, he told his story. He was first taken by St. Nicholas to the places of punishment, one of which was a stinking body of water. Sinners were punished according to their sins. Then he was shown the "Heavenly Jerusalem and Christ on the cross." He was taught that one may work one's way through punishment to reward, no matter what the sin.

A third vision came from a laborer named Thurkill and was reported in 1206. During his experience, St. Julian guided him to a purgatorial fire, a cold and salty lake, and a bridge with thorns and stakes. He saw St. Paul and a devil weighing people on a scale to decide the fate of their souls. The place of punishment was like a theatre where sinners re-enacted their sins. The theater seats inflicted pain on the audience as the sinners were tortured by devils. Thurkill met several saints, two apostles, an archangel, and Adam during his vision. Heaven was a giant church called the Congregation of the Saints. When he returned to his body, Thurkill was reluctant to tell his story until he experienced a second vision that instructed him to tell others so that they might learn from his experience.

Emanuel Swedenborg

Emanuel Swedenborg was a Swedish scientist, philosopher, theologian, and mystic who was born in the late 1600s. In 1741, at the age of 53, he began to have dreams and visions. His experiences led to a "spiritual awakening." He claimed that Jesus Christ had appointed him to write

The Heavenly Doctrine to reform Christianity. Swedenborg said his spiritual eyes had been opened so he could freely visit heaven and hell and speak with a variety of spirits. During last 28 years of his life, Swedenborg published 18 theological works.

Although Swedenborg wrote about a "New Church" that would be based on his writings, he never tried to establish an organized religion. His writings focused on one universal church based on love and charity, rather than multiple churches with different doctrines. His new ideas did not sit well with the leading church at the time. When two men began to promote his vision, they were arrested for heresy. A trial began in 1768, and in 1770 a royal ordinance declared that his writings should not be taught. When his supporters were ordered to stop using his teachings, Swedenborg begged the king for mercy. The case was dropped in 1778. In 1787, 15 years after Swedenborg's death, the New Church movement was founded in England. Soon after, his teachings were brought to the United States by missionaries, one of whom was John Chapman (Johnny Appleseed). Today, the Swedenborg Foundation is a nonprofit organization that helps people learn about his teachings.

Swedenborg referred to the passing of the spirit as the awakening. He said, "I have not only been told how the awakening happens; I have been shown by firsthand experience. The actual experience happened to me so that I could have a full knowledge of how it occurs."

He explained the process of passing over to the other side. "We are only separated from the physical nature that was useful to us in the world. The essential person is actually still alive. I say that the essential person is still alive because we are not people because of our bodies but because of our spirits." He saw death as a resurrection or a continuance of life.

Swedenborg said that he was first contacted by angels. They sat near his head, silent, sharing their thoughts with him in a telepathic way. They kept watch over him, waiting for his body to breathe its last breath. "When heavenly angels are with people who have been awakened, they do not leave them, because they love everyone." At the same time, he felt a tug, like his spirit was being pulled from his body.

He spoke about angels rolling back a covering from his left eye toward the center of his nose so that he could see. "I could see a kind of clear but dim light like the light we see through our eyelids when we are first waking up. It seemed to me as though this clear, dim light had a heavenly color to it," he said. After that, something rolled back from his face.

According to Swedenburg, angels do everything for newly arrived spirits. They inform them, at least to the extent that they can grasp, about the realities of their new life. If newly arrived spirits do not want to be taught, they are free to move on. Other good spirits accompany them and do all they can for them. If the departed was not a good person in the physical world and wanted to get away from these good intentioned spirits, they were free to go. They may search until they find the company of people more attuned to their own spiritual development. Swedenborg explained that some spirits "remarkable as it may sound, then lead the same kind of life they had led in the world." This first stage of their life on the other side does not last more than a few days. After that, the spirit travels from one state into another until finally it arrives either in heaven or in hell.

In his book, *Heaven and Hell,* Swedenborg said he had spent twenty-seven years of his life in almost constant communication with angels and devils in the spiritual realms. He described the nature of heaven and hell, and his conversations with the inhabitants of those realms.

Swedenborg wrote:

> *It is recognized that non-Christians live lives that are just as moral as the lives of Christians—many of them, in fact, live more moral lives. A moral life may be lived either to satisfy the Divine or to satisfy people in this world. A moral life that is lived to satisfy the Divine is a spiritual life. The two look alike in outward form, but inwardly they are totally different. One saves us, the other does not. This is because if we live a moral life to satisfy the Divine we are being led by the Divine; while if we live a moral life to satisfy people in this world, we are being led by ourselves.*
>
> *People who have intended and loved what is evil in the world intend and love what is evil in the other life, and then they no longer allow themselves to be led away from it. This is why people who are absorbed in evil are connected to hell and actually are there in spirit; and after death they crave above all to be where their evil is. So after death, it is we, not the Lord, who cast ourselves into hell.*
>
> *Seen in its own right, a loving person's inner spirit is an angel of heaven, and while we are living in the body we are also in the company of angels, even though we are not aware of it; and once we are released from the body we join them.*

Andrew Jackson Davis

Andrew Jackson Davis was born in Blooming Grove, New York, a small town along the Hudson River. Because his family moved frequently, Davis received little education. He eventually apprenticed with a shoemaker for two years and would have run a typical business if he hadn't attended a lecture on Animal Magnetism given by Dr. J.S. Grimes in Poughkeepsie, New York.

Davis tried the hypnotic technique taught by Grimes but didn't have any success with it until a local tailor, William Livingston, helped him reach a trance state. While in this state, he claimed to understand the laws of the universe. He diagnosed and prescribed cures for individuals and explained spiritual and metaphysical phenomena. He soon became known as the "Poughkeepsie Seer."

Even though he only had 5 months of formal schooling and claimed to have read only a handful of books, he dictated a book in 1846 entitled *Principles of Nature*. It was published a year later when he was only 21. He eventually published over 30 books with topics that included the seven planes of existence, health, science, philosophy, and education. In the *Great Harmonia*, he refers to the process of evolution nine years before Darwin published his book, *On the Origin of Species*. He also predicted the existence of the planets Neptune and Pluto before their discovery.

Davis predicted the coming of Spiritualism in *Principles of Nature* (1847), saying, "And the world will hail with delight the ushering in of that era when the interiors of men will be opened, and the spiritual communion will be established." The Fox family would have their encounter with their famous rapping spirit just a year later. Several sources during the time mentioned that Davis met with President

Abraham Lincoln, both at the White House and in Davis's home in New York. Davis eventually obtained a medical degree and moved to Boston where he opened a bookshop and prescribed herbal remedies. He died in 1910.

Summerland was a term used by Spiritualists to refer to heaven or the spirit world. It first appeared in 1845 in Davis's published automatic writings (writing without consciously controlling the process). One of several concentric spherical planes surrounding the earth, Summerland was described as the residence of spirits of goodwill. A basic principle of Spiritualism is that "like attracts like." Deceased people, who were attuned to beauty, love, and learning are drawn together in the afterworld. Summerland was a product of the minds of its inhabitants. Davis's communications described the deceased as living a harmonious, quasi-physical life amid beautiful houses, lecture halls, music, gardens, meadows, trees, lakes, streams, and animals. A spirit residing in this sphere could also communicate with the living for their benefit; likewise, living people while out-of-body may upon occasion visit the Summerland. It must be noted that some of these ideas appear in the earlier writings of Emanuel Swedenborg. Davis reported that while he was entranced, Swedenborg dictated the material to him.

Edgar Cayce

Edgar Cayce was born in 1877 on a farm near Hopkinsville, Kentucky, one of Leslie and Carrie Cayce's six children. When he was ten, he became engrossed in the *Bible* and read it through over a dozen times. In 1889, while reading the *Bible* in the woods, he had a vision of a woman with wings who asked him what he wanted most of all. He said he wanted to help people. His first thought was to be a Christian missionary. She told him he would be able to help others while he was asleep.

Edgar only attended school until the 9th grade. During that time, he discovered he could learn his lessons by sleeping on his books. As a member of the Disciples of Christ church, he read through the *Bible* once a year and began teaching Sunday school. At that time, he said he could also see auras, speak to angels, and hear the voices of his departed family. He agonized over his abilities, not sure if they were the work of God or some evil force.

At the age of twenty-one, he developed "chronic throat paralysis" which resulted in the loss of his voice. When doctors were unable to find a cause for his condition, Edgar asked for help from a hypnotist named Hart. They found that when Edgar was in a hypnotic trance he could speak normally, but when he woke up his voice disappeared. Hart was replaced by Al Layne who eventually asked Cayce what was needed to cure his voice. "Increase blood circulation to the throat" was the answer. Cayce followed the recommendation. Although his problem would reappear sporadically, he was essentially cured.

Cayce and Layne soon discovered that Edgar could recommend cures for others while in trance. For most of his adult life, he provided answers for a great variety of questions while in his induced sleep state. His responses to both people who appeared in person and to those who sent letters were called "readings." His words were transcribed by Al Layne, his wife, Gertrude Evans, and later by his secretary, Gladys Davis Turner. He would eventually give over 14,000 readings on more than 10,000 different topics for people from all over the world.

For many years his readings focused on medical problems. Eventually the scope of topics expanded to include meditation, dreams, reincarnation, and prophecy. Many of these topics were at odds with his

Christian beliefs, and for a time Edgar did not believe the stenographers were accurately copying what he said while in the trace state. Eventually, he came to terms with the topics he discussed while in trance.

His statements about death and the afterlife included the following:

> *Death is as but the beginning of another form of phenomenized force in the earth's plane, and may not be understood by the third-dimension mind from third dimension analysis, but must be seen from that fourth-dimension force as may be experienced by an entity gaining the access to same, by development in the physical plane through the mental processes of an entity (Reading 136-18)*
>
> *Life and its expressions are one. Each soul or entity will and does return, or cycle, as does nature in its manifestations about man; thus leaving, making or presenting—as it were—those infallible, indelible truths that it—Life—is continuous. And though there may be a few short years in this or that experience, they are one; the soul, the inner self being purified, being lifted up, that it may be one with that first cause, that first purpose for its coming into existence. Reading 938-1*

In Reading 5749-3, he discussed the phases of spiritual development and reincarnation.

> *When there was in the beginning a man's advent into the plane known as earth, and it became a living soul, amenable to the laws that govern the plane itself as presented, the Son of man entered earth as the first*

man. Hence the son of man, the son of God, the son of the first Cause, making manifest in a material body.

This was not the first spiritual influence, spiritual body spiritual manifestation in the earth, but the first man—flesh and blood; the first carnal house, the first amenable body to the laws of the plane in its position in the universe.

For, the earth is only an atom in the universe of worlds!

And man's development began through the laws of the generations in the earth, thus development, retardment, or alterations in those positions in a material plane."

Later he said:

Passing from the material consciousness to a spiritual or cosmic, or outer consciousness, oft does an entity or being not become conscious of that about it; much in the same manner as an entity born into the material plane only becomes conscious gradually of that designated as time and space for the material or third dimensional plane. In the passage the entity becomes conscious, or the recognition of being in a fourth or higher dimensional plane takes place, much in the same way as the consciousness is gained in the material.

For, as we have given, that we see manifested in the material plane is but a shadow of that in the spiritual plane.

These readings continue to be researched and written about over half a century after his death and are available to students, writers, researchers, medical professionals, and Association of Research and Enlightenment (A.R.E.) members the world over.

SECTION VI
MODERN ACCOUNTS

Chapter 18
Near-Death Experiences

Since the 1970s when Raymond Moody coined the term Near-Death Experience, there has been growing interested in the phenomena, both by the general public and by the scientific community. NDEs have been recorded around the world, but fifty years after Moody's first incites were published, we still don't know if those at death's door are seeing an afterlife, or if they are hallucinating. This chapter discusses some published cases and the research happening today.

Eben Alexander Case

Eben Alexander graduated from the University of North Carolina at Chapel Hill in 1976 with a degree in chemistry and Duke University with a medical degree. His residency training was completed at Duke, Massachusetts General Hospital and Harvard. With a main interest in neuroendocrinology, he focused on the nervous system and hormones. Specifically, he is interested in a syndrome called cerebral vasospasm, how blood vessels react when there is bleeding from an aneurysm. Dr. Alexander also completed a fellowship in the UK and spent 15 years on the faculty of Harvard Medical school as an associate professor of surgery.

In 2008, at the age of 54, Dr. Alexander suffered with what he thought was a typical respiratory virus. He, his wife and son had mild symptoms, but his symptoms did not abate. One day he woke up with severe back pain. Not long after he slipped into unconsciousness. His wife called 911 and they rushed him to the hospital. By then, he was convulsing violently and groaning. A spinal tap, taking a sample of spinal fluid to test, revealed that he had bacterial meningitis, a swelling of the membranes of the spine and skull.

The news was not good. The medical team loaded his body with three powerful antibiotics and sent him to the intensive care unit. He was sedated and a breathing tube was inserted to keep him alive while he fought the infection. As the days passed, he was at serious risk of dying and the longer he stayed in a coma, the greater the chance that he would end up in a permanent vegetative state.

During his weeklong coma, Dr. Alexander described his experience. "I didn't have a body—not one that I was aware of anyway. I was simply...there, in this place of pulsing, pounding darkness. At the time it was going on, I didn't know this word. In fact, I didn't know any words at all. The words used here registered much later, when, back to the world, I was writing down my recollections. Language, emotion, logic: these were all gone...." He found himself in earthy darkness, with faces and animals bubbling up around him.

At a certain point, filaments of pure white light surrounded him, and he rose into a new world. It was like earth, but "A beautiful, incredible dream world...." A winged woman joined him and communicated to him without speaking. She told him he would see many beautiful things but would eventually have to go back.

He wrote in his book, *Proof of Heaven: A Neurosurgeon's Journey into the Afterlife,* "The unconditional love and acceptance that I experienced on my journey is the single most important discovery I have ever made, or will ever make, and as hard as I know it's going to be to unpack the other lessons I learned while there, I also know in my heart that sharing this very basic message—one so simple that most children readily accept it—is the most important task I have."

Dr. Alexander experienced an event much like other NDE survivors. He wrote: "The place I went to was real. Real in a way that makes the life we're living here and now completely dreamlike by comparison." But is his experience proof that consciousness survives beyond death, proof of heaven as his book proclaims? Like all other experiences, his is an antidotal report. There is documentation that critical portions of his brain were not functioning for most of the seven-day period he was in a coma. But when did this experience occur? Was it during the seven days, or did it occur in those minutes his brain was awakening, and he was coming out of the coma? There is no way to tell because there was no way to document when the events were happening.

Raymond Moody

The person most closely associated with Near Death Experiences is the man who coined the term, Dr. Raymond Moody. Moody was born in Georgia in 1944. He earned a BA, MA and PhD in the 1960s from the University of Virginia in philosophy. He also received a PhD in psychology from the University of West Georgia and an MD from the Medical College of Georgia in the 1970s. After working as a forensic psychiatrist in a maximum-security hospital in Georgia, he was appointed chair in Consciousness Studies at the University of Nevada, Las Vegas. Moody's first exposure to near-death experiences (NDEs) was an account told to him by Dr. George Ritchie in 1965. Richie later recounted the experience in his book *Return from Tomorrow* published in 1978.

Although near death experiences have been recorded as far back as written language, they were never documented in any sort of meaningful way. Moody was the first to record these experiences and try to quantify similarities and differences. These are anecdotal accounts and the process that causes these experiences is still unknown. Despite that, Moody was able to develop a list of experiences that were common to all observers.

Common traits that have been reported are:

1. An awareness of being dead
2. A sense of peace and well-being and intense feeling of unconditional love and acceptance
3. An out-of-body experience, seeing one's body from the outside.
4. A sense of moving up through a passageway, staircase or dark tunnel
5. A powerful light or being of light which communicates telepathically.
6. Reunited with loved ones
7. Life review
8. Deciding to return to their body
9. Finding that they are back in their body

Not everyone who has a near-death experience reports every event listed here. Some recall only an out-of-body experience. Others get as far as the dark passageway before turning back, while others proceed all the way into a bright light.

During a typical NDE, the patient will describe how they leave their body and rise upward. They may witness the resuscitation procedure going on below, not realizing they are looking at their own body at first. Subsequently, they pass through a dark enclosure, sometimes described as a tunnel, until they reach a brilliant white light. Most say that words are inadequate to describe what happens next. The bright light brings them comfort, joy, peace and love. They also may see deceased loved ones who are there to welcome them, and they experience events of their lives in a full-color panoramic way.

The patients return to their bodies in different ways. Some find themselves back in their hospital beds with no memory of a transition. Others say they were told they must return for a variety of reasons. Either way, the patients admit that their NDE profoundly changed their lives. They no longer fear death. Their old pursuits become less important, and they focus on learning to love mankind.

Experiencing NDEs are not determined by a person's age and do not seem related to their religious beliefs. Many people with no interest or background in religion report powerful NDEs. They are also not related to specific medical conditions. Infections, trauma, cardiac arrest, complications of childbirth and other conditions have all been reported at the time of the experience. Those who have had NDEs insist they are not like dreams. In comparison, dreams seem less real than ordinary waking reality and NDEs seem hyper-real.

Moody's thorough data collection has described thousands of NDE experiences. But these are anecdotal stories. They may be real, or they may be complex seizures or brain anoxia (loss of oxygen) during traumatic events, as some have suggested. Without additional research we will never know.

Bruce Greyson

Bruce Greyson is Chester F. Carlson Professor Emeritus of Psychiatry and Neurobehavioral Sciences, Professor of Psychiatric Medicine in the Department of Psychiatric Medicine, and the former director of The Division of Perceptual Studies (DOPS) at the University of Virginia. Grey states that he was raised a materialist, believing that only the material world existed and that everything experienced by the mind was due to the functioning of the brain.

Greyson had his first brush with an NDE patient during his first medical internship. He was asked to examine a woman who had overdosed. When he reached the emergency room, he found she was unconscious. Unable to speak with her, he talked to her roommate in a waiting room down the hall.

This might have just been a normal "this patient is imagining things" event if he hadn't spilled spaghetti sauce on his tie and buttoned up his lab coat to cover it up. While with the patient, his lab coat was closed. When he spoke with the friend, he unbuttoned the coat because it was hot in the room. When he visited the patient the next morning and started to introduce himself, she told him she remembered him from the previous night. When he stated that she was unconscious, she told him she had been there listening to the conversation he had had with her roommate and about the spaghetti stain she saw on his tie.

Greyson never spoke of that first experience, but during his psychiatric training in the 1970s, he was confronted by patients who spoke of leaving their bodies and seeing and hearing things from an out-of-body perspective. To understand what was happening, Greyson along with others, decided to build upon the research of Dr. Moody.

He developed the Greyson Scale to measure aspects of NDEs which has been widely used.

After his retirement in 2015, to maintain the continuity of the ongoing NDE research, Greyson has continued to collaborate with scientific colleagues from all around the world. Dr. Marieta Pehlivanova became a member of the DOPS staff in 2017. Greyson has been mentoring and collaborating with her on a variety of research projects recording the immediate and long-lasting effects of near-death experiences. They are confident that NDE experiences cannot be dismissed as dream states or hallucinations but have conducted no brain studies or scientific tests to determine their cause.

Blind NDE Experiences

If NDEs are complex hallucinations of the brain, one would expect that blind people would only report on experiences that they could perceive in waking life. One would not predict that they could "see" their body during an out-of-body experience, or describe colors, bright light, and objects that they wouldn't normally see.

Kenneth Ring and Sharon Cooper interviewed blind participants for a research paper that was published in 1997. They questioned 31 blind respondents in total. The study sought to address three main questions: (1) do blind individuals have near-death experiences (NDEs) and are they the same as or different from those of sighted persons; (2) do blind persons ever claim to see during NDEs and out-of-body experiences (OBEs); and (3) were their claims corroborated by independent evidence.

Ring and Cooper screened 46 people and included 31 in their interviews. There were 20 women and 11 men, ages 22-70 years. All were Caucasian and most Christian. Twenty-one had NDEs and 10 were people who had one of more Out of Body Experiences (OBEs) only. NDEs were associated with traumatic events and OBEs occurred during trauma or while the person was relaxed state. Fourteen of the people were blind from birth, most having no light perception at all. Eleven had lost their sight after the age of five. The remaining six had severely impaired vision. Of those who had NDEs, 10 were blind from birth, nine were blind after the age of five and, 2 were severely impaired. Of those who had OBEs, 4 were blind from birth, 2 were blind after age five and 4 were severely impaired.

Previous research by D.D. Kirkley found that congenitally blind people and those blinded before the age of five have no visual images in their dreams. The blind do not have any visual impressions. Hearing tends to be the primary sense experienced in dreams along with tactile feelings. One would expect that NDE and OBE experiences would be similar for the blind. That was not the case.

Ring and Cooper included some people who were blind after the age of five or who were severely impaired, but 9 out of the 14 people blind from birth reported sight. "Overall, 80 percent of our respondents reported these claims, most of them in the language of unhesitating declaration, even when they had been surprised, or even stunned, by the unexpected discovery that they could in fact see." One man found himself in a large library and saw "thousands and millions and millions of books, as far as you could see." He stated that he saw them visually and clearly.

Near Death Experiences offer some insight into the existence of heaven. But we still don't know if these are actual experiences or creations of the mind. Out of body experiences may give us a clue about survival of the consciousness after death.

Chapter 19
Out of Body Experiences

An Out of Body Experience (OBE) is the sensation that the self or consciousness is leaving the body. During these events, people view the world around them from an external vantage point instead of the usual sense of self from inside their physical body. OBEs are associated with near-death experiences, but they can also occur during sleep and may be triggered by medical conditions like migraines and epilepsy. OBEs typically happen without warning and usually don't last for very long. According to accounts, they generally involve: a feeling of floating outside the body, an altered perception, such as looking down from a height, and a sense that what's happening is very real.

Some estimates suggest that around ten percent of people have experienced an OBE. Anecdotal stories report individuals, usually patients, leaving their body and seeing things from above themselves, hearing what is being said in the room, and even leaving the room and witnessing activity in other areas. There is plenty of debate over whether the sensations and perceptions happen physically or are a sort of hallucinatory experience. Reports have been published over the years, some with very specific accurate details about events that happened during surgical procedures or while clinically dead. But what about scientific research? Is there a way to test whether OBEs are real or imaginary? Maria is one of thousands who have reported OBEs.

Shoe on the Sixth Floor

Harborview Medical Center is a massive, multi-story medical complex in the heart of Seattle. Being a comprehensive healthcare facility, they provide specialized care for a broad spectrum of patients from the Pacific Northwest, including the vulnerable residents of King County. It was one of those vulnerable residents, a migrant woman named Maria, who was brought to the Emergency and Trauma Center entrance. When she was whisked into doors beneath red lights spelling emergency room, she was having a massive heart attack. What happened next has become somewhat of an urban legend.

To understand Maria's story, it is important to understand the story of another woman, Kimberly C. Sharp. Sharp is a retired Clinical assistant professor of social work at the University of Washington. On a spring day in 1970, outside an office in Shawnee Mission, Kansas, the electric impulse that regulated her heartbeat misfired. She was only 22 years old, and her heart suddenly stopped. She had no pulse and wasn't breathing. Her helpless father watched as strangers worked on her and tried to revive her.

While Sharp was clinically dead, she moved into a place of brilliant golden light and warm comfort. The experience changed her, and she felt she understood the true meaning of life and death--service. She moved to Washington to devote herself to caring for others. With a master's degree in social work from the University of Washington, she began a career in medical social work. She found herself in direct contact not only with dying people, but with people who almost died and came back.

When Sharp published stories of people's experiences in her book, *After the Light: What I Discovered on the Other Side of Life that Can*

Change Your World in 1995, Maria's story was in the first chapter. Sharp first met Maria when she was a patient in the cardiac care unit. Sharp's job was to reassure her that they would contact her family and find the money to pay for her hospitalization. Maria's English was poor, and Sharp's Spanish wasn't much better, but they were able to communicate.

After three days in the cardiac unit, Maria was looking better, and Sharp had grown fond of her. That's when Maria had her second cardiac arrest. Monitors blared and the resuscitation team ran to her room and began CRP. It only took a couple of electrical shocks to restart Maria's heart. She was unconscious but in stable condition.

Several hours after her emergency, Maria gained consciousness and was breathing on her own. Sharp was getting ready to leave for home when one of the nurses called her. Maria was so agitated that the nurse was afraid she would have another cardiac event. Sharp went to visit her to try to calm her. Maria was flailing her arms and ranting in Spanish when she arrived. Sharp tried to comfort her and Maria's excitement turned to sobs.

When Maria finally calmed down, she told Sharp about floating to the ceiling and watching people work on her body during her cardiac arrest. She described all the paper from the electrocardiogram machine that had been kicked around the room. But the story didn't end there. Maria said she left the hospital room and drifted outside the building where she could see the emergency sign, all the vehicles arriving at the entrance, and the automatic doors opening. It was during her out-of-body experience that Maria spotted a man's dark blue tennis shoe on a window ledge of the hospital, about three stories above the ground. She described the shoe in detail. She knew she wasn't crazy and asked Sharp to look for it.

Sharp doubted that she was going to find the shoe. She checked ledges for the rooms on the east side and north side of the hospital and found nothing. While in the fourth room on the west side of the building, she pressed her face against the window glass and saw the shoe. It was well-worn, the toe was scuffed, and a lace was tucked under the shoe's heel, just as Maria had described. She carefully opened the window and retrieved the shoe. The day Maria was discharged, Sharp shared her own near-death experience with her.

After that event, Sharp became the founder of the Seattle International Association of Near-Death Studies in 1982. She was also founder of the Department of Social Work at the Fred Hutchison Cancer Research Center, the world's first bone marrow transplant center. As a pioneer in the field of critical care social work, Sharp is an international conference and workshop speaker, a consultant to news media, and published in many textbooks, journals, and magazines. She worked for two decades as a clinical assistant professor at the University of Washington and co-taught a Terminal Illness Seminar at the University of Washington, School of Medicine.

Pam Reynolds Case

A second compelling case is that of Pam Reynolds. Reynolds was born in 1956 in Atlanta, Georgia and worked as a singer-songwriter until she began experiencing symptoms of dizziness, loss of speech and difficulty moving. Her physician referred her to a neurologist, and a CAT scan revealed a large aneurysm in her brain. Because of the location of the aneurysm, Reynolds was given little hope of surviving surgery but wanted to go ahead with it.

Robert F. Spetzler, a neurosurgeon of the Barrow Neurological Institute in Phoenix, Arizona, opted to perform a rarely used procedure known as a standstill operation. During this process, Reynolds' body temperature was lowered to 60 °F, her breathing and heartbeat were stopped, and the blood was drained from her head. Her eyes were closed with tape. Ear plug speakers which emitted loud clicks every few seconds were used in conjunction with an EEG to ensure that her brain was unresponsive. The surgery lasted about seven hours.

Reynolds' NDE began with an OBE. She said she heard a sound like a "natural D" that pulled her out of her body. She floated above the operating room and watched the doctors perform the surgery. During that time, she claimed that she felt "more aware than normal" and her vision was more focused and clearer. She also described seeing a surgical saw that looked like an electric toothbrush (an accurate description of the saw). She heard conversations between operating room staff, even though the ear plug speakers were making loud clicking noises.

After the OBE, she noticed a presence and was pulled towards a light. She saw figures in the light, including her grandmother, an uncle, additional deceased relatives, and people she couldn't identify. She didn't want to go back to her body but was told that she didn't have a choice. When she returned to her body, she said it felt ice cold.

One would think that the Pam Reynolds case is compelling proof of consciousness leaving the body. She was monitored during the entire operation. There was no brain-wave activity and no blood flowing in her brain during the surgery when she reported her observations. Yet, controversy remains.

Anesthesiologist Gerald M. Woerlee, who has many years of clinical experience, believes that Reynolds may have experienced anesthesia awareness. Although rare (0.1-0.2 %), some patients are not completely anesthetized during medical procedures. They may report everything from vague, dream-like states to being fully awake, immobilized, and in pain during procedures. The NDE could have occurred while Reynolds was under general anesthesia before the brain was cooled down. He was not impressed by her description of the saw and believes the remainder of the experience came to her during recovery. He also believed that Reynolds could hear during periods of conscious awareness while under the influence of general anesthesia drugs.

Chris Carter PPE, MA, teaches mathematics and theory of knowledge in the International Baccalaureate program based in Geneva, Switzerland. He responded to Woerlee's article about Reynold's experience in the Journal of Near Death Studies, arguing that Woerlee's explanation was unsupported by the documented facts of the case. In his long rebuttal article, he takes the reader through the steps of the surgery. He insists that Reynolds was not hearing anything while under anesthesia and invited Woerlee to accompany him to the Barrow Neurological Institute in 2011 to participate in an empirical test under the exact auditory conditions Reynolds experienced.

It appears that Woerlee is a militant skeptic. I will not step into that fray with this book, but it's important to understand the difference between hypothesis testing and skepticism. All scientists are skeptical. That is why they question how and why the universe works. They devise tests that get rid of their biases to answer their questions. On the other hand, some skeptics set up a bias that they then try to prove. But they do not use hypothesis testing to do so. Instead, they cherry pick information from sources to support their arguments.

Scientists have begun to quantify OBE data, just as Moody did with NDEs. Unfortunately, there is no way we can determine if what a patient has witnessed is a hallucination or not just by interviewing them after the fact. Science relies on testing hypotheses to determine if something is occurring. Is it possible to set up a controlled research experiment to determine if a person's consciousness leaves their body after death? It would be unethical to cause the death of a person to study the phenomena, but there are other ways to collect data.

Chapter 20
Ghost Hunting

Ghost hunting is a blend of modern technology and the belief in spirit communication. As part of Spiritualism in the 19th and 20th centuries, ectoplasm, knocks, apparitions, apports, and later photographs were considered physical manifestations of spirits. Today there are hundreds of websites on the Internet dedicated to the investigation of ghostly spiritual phenomena. These promote a large diversity of methods and approaches that attempt to analyze paranormal happenings. The sites publish photographs, sound recordings, and personal accounts. Advice is available and ghost hunting kits are for sale.

Television shows are also popular, with over a dozen ghost hunting series and many more paranormal shows. One of the first shows was *Ghost Hunters* which ran from 2004 -2016 on Syfy, then went to A&E and Discovery + in 2021. Jason Hawes and Grant Wilson of The Atlantic Paranormal Society (TAPS) began as plumbers by day and ghost hunters by night. They appeared sincere in their search for paranormal happenings, but many of the other series that followed did not seem to be as thorough in their methodology. And all the shows appear to be dramatized for the audience.

Modern ghost hunters use a variety of equipment to detect and record encounters with spirits. These include Geiger counters, Electromagnetic Field (EMF) detectors, ion detectors, and regular and infrared cameras. Everything seems very technical and scientific, but are they really contacting spirits in the afterlife or is something else happening?

Electronic Voice Phenomenon

Attila von Szalay, an American photographer, was one of the first to try to produce audio recordings as well as photographs of spirits. He began in 1941 with 78 rpm records, but it wasn't until 1956, using reel-to-reel tape that he had any success. With the help of Raymond Bayless, von Szalay tested his custom-made recorder, a microphone in an insulated cabinet connected to an external recording device and speaker. After conducting several recording sessions, they discovered many sounds on the tape that were not heard at the time of recording. They believed these sounds to be spirits. The messages included "This is G!", "Hot dog, Art!", and "Merry Christmas and Happy New Year to you all."

Konstantin Raudive, a Latvian psychologist who had taught at Uppsala University, Sweden, worked with filmmaker Friedrich Jürgenson, making over 100,000 recordings which they described as containing communications with discarnate people. Some of the recordings were conducted in a radio frequency-screened laboratory. Raudive invited listeners to hear and interpret the recordings. He believed that the clarity of the voices implied that they could not be explained by normal means. He published his first book, *Breakthrough: An Amazing Experiment in Electronic Communication with the Dead* in 1968 (English in 1971). Since then, spirit communications have been recorded using audio tape, radio, diodes, television, and digital audio recorders.

A major criticism of this technique is that the recorded sounds may be more imagined than actual voices. Tapes and audio recordings produce a background noise or hiss. The listener must distinguish a voice or words from this noise. The problem is, just like humans are prone to Pareidolia (seeing faces in many things), it's possible to believe one hears words in abnormal sounds. The term for this is Auditory Pareidolia. It is a situation created when the brain incorrectly interprets random patterns as being something familiar. The famous psychologist, B. F. Skinner, in 1936 played nonsense sequence of vowels to subjects after they were instructed to "listen with care." He discovered that they believed they heard words where none existed.

Verbal Transformation is another effect which can happen when listening to audio recordings repeatedly. Richard M. Warren, in 1968, researched the phenomenon and found that his subjects would give several different meanings to a repeated word. When 18 people heard the word "tress" 429 times over a 3-minute period, they would start to hear different words in the recording. One subject heard a total of 17 different words during the sequence.

Because EVPs could be the result in an investigator interpreting random noise as a human voice, there needs to be strict scientific controls to determine if the voice or words have actually been recorded, or if they are imagined. Imants Baruss decided to put the phenomenon to the test. With a background in engineering science from the University of Toronto, an interdisciplinary BSc, MSc in mathematics from the University of Calgary, and a PhD in psychology from the University of Regina, he conducted a study of EVPs that was published in 2001.

Baruss's experiment used a cassette recorder, microphone, and two radios placed at right angles in a private room at King's College. Research assistants were used to simulate interaction with spirits while taping the output from the two radios tuned between stations. There were 81 sessions that averaged approximately 45 minutes in length for a total of about 60 hours of recording. Baruss noted that "while there are some apparent voices and interesting noises upon playback, none of these is sufficiently distinctive to merit being considered anomalous."

Although Baruss's experiment failed to produce any concrete evidence, it was the first scientific attempt to test EVPs. Failure may be due to lack of contact with the spirits. The experiments were held at the university, not at a site of reported ghostly encounters. They may have needed a person with psychic ability present to act as a catalyst to start communication. The equipment may also have been inferior. In any case, more experiments should be conducted.

Ghost Photography

Spirit photography first began in 1839 using the daguerreotype process. To make the image, a polished sheet of silver-plated copper was treated with mercury fumes that made its surface sensitive to light. After it was exposed, it was immersed in a chemical liquid to set the image. The image was on a mirror-like silver surface that would appear positive or negative, depending on the angle at which it is viewed. Because they require a long exposure time, people or objects passing through the area being photographed could leave a faint image. In 1856, Sir David Brewster discovered that these effects could be used to create ghostly pictures, and the London Stereoscopic Company used his idea to create a series of images called The Ghost in the Stereoscope. So, the first spirit photographs were purposeful fakes.

When the glass-plate negative processes began around 1859, exposure plates could be reused, and previous images could remain visible. As cameras prices fell, spirit photos could be created by overlapping previous images with new, and spirit photography became widespread. One of the most famous photographers was William

Hope. He was tested by members of the Society of Psychical Research in 1922. They marked negatives with x-rays to make sure Hope was using their negatives. When the negatives were returned with photographs of spirits, the researchers found that the marks were not there. Hope had switched negatives and was found to be a fraud.

It is difficult to tell if any of the old spirit photographs are real. Many were of poor quality and looked like cut-outs of faces overlapping with the image of a person. Spirit photographs were not taken in a controlled setting, but in a studio or séance and developed outside, which gave ample time to create fake photos.

What about modern ghost photography? Ghostly photographs are usually taken at the location of the haunting and technology has changed dramatically over the years. We have gone from film cameras, to Polaroids, to digital cameras that create still shots and videos with sound. Today, most people have cellphones with cameras. Unfortunately, camera apps can now place images of ghosts, aliens, and monsters into actual photographs. In 2014, there were over 250 ghost applications for Android phones. We also have Artificial Intelligence creating incredible fakes in both still photographs and motion pictures. The ability to fake a photo has improved to such an extent that real photos would be difficult to distinguish from simulations. This makes the need for controlled scientific tests conditions more imperative if we want to determine if ghost photography is real.

Electromagnetic Frequency Meters

Electromagnetic fields are created by nature and human-made sources. From electrical charges created by thunderstorms to X-Rays, TV antennas, electrical wiring, and electrical appliances, it is not uncommon to encounter them. An electromagnetic field meter (EMF) is a device that is designed to detect and measure electromagnetic radiation produced by the field. Traditionally it's been used in a variety of settings, including industrial, scientific, and medical fields, as well as in home settings. They work by detecting electric and magnetic fields and translating the information into a readable form, such as a numerical value or an audible alert. Some EMF meters detect only a narrow range of frequencies, while others detect a broad range of frequencies.

EMF meters are one of the most used tools for ghost hunting. The idea is that ghosts are made up of energy, and because of that they can manipulate or emit electromagnetic fields. The assumption is that the presence of this radiation can indicate a visitation from an otherworldly presence. Because they're handy, relatively inexpensive, and easy to use, EMF meters are popular.

There are often ordinary explanations for what EMF detectors pick up and electromagnetic radiation can be found virtually everywhere. Non-paranormal sources such as faulty wiring, microwave towers, sunspot activity and even electronic equipment such as walkie-talkies, TV cameras and wiring used by ghost hunters can create EMF spikes.

Like the physical phenomena produced at the turn of the 20th Century, most ghost hunting assumptions have not been tested in controlled situations. How can EVPs be analyzed when the human brain tends to create meaning from nonsensical sounds? Maybe AI technology could be used in the future. Controlled experiments need to be conducted with photography so that fraud can be eliminated. And new methods need to be tested to make EMF readings a reliable source of information.

Noah Leigh is a trained scientist (BA Biology, MS in Epidemiology and Cell Biology). He uses his scientific training to run Paranormal

Investigators of Milwaukee (PIM's) as much as possible as a controlled experiment. He investigates with a skeptical approach, believing that that whatever can't be debunked may be real evidence of the paranormal. He uses careful measurements and observations to ensure the validity of any data recorded. His Paranormal 101 section of PIM's website offers insight into paranormal phenomena. He has not conducted controlled experiments as of yet.

SECTION VII
DOCUMENTATION AND TESTING

Chapter 21
Testing Mediums

Scole Experiment

Although Spiritualism lost its popularity by the mid-1900s, organizations and mediums continued their quest to demonstrate the survival of the soul after death. One such example is the Scole Experiment. It took place between 1993 and 1998 under the direction of researchers and mediums Robin and Sandra Foy and Alan and Diana Bennett in the cellar of Foy's 17th century farmhouse located in the village of Scole in northeast England. The group conducted many experiments that involved video and still cameras, camera film (not removed from the can), and voice recorders. They received instructions from the spirit world to enhance their ability to communicate and tried to control each experiment to remove bias and trickery.

An example of one session took place in November 1996, their guests included researchers Montague Keen, David Fontana (Professor of Psychology) and Arthur Ellison (Professor Emeritus of Electrical Engineering). An anonymous spirit scientist instructed them through one of the mediums to build a device. It was to incorporate a piece of the metallic element, Germanium, which would be put under a small amount of pressure.

The spirit discussed technical issues with Professor Ellison. The Germanium device was a receiver which would be attached to a battery-run amplifier and tape recorder. The device did not receive radio signals but worked on an unknown principle that would help them communicate with the spirit world. There was no microphone attached so only communication from the spirit world could be recorded, not the Scole participants.

At first, when they played back the recorder, all they heard was white noise and occasionally faint voices. They received more information in the form of a diagram that appeared on film (sealed in its original canister) once it was developed. It showed an additional two coils added to the device. Three letters were inscribed near the diagram, T.A.E.

Professor Ellison was brought in to modify the machine. On January 21st, 1997, they proceeded with the experiment. At first there was only white noise. Then they heard a faint voice calling, "Hello." There continued to be problems, and the amplifier appeared to have been turned off at some point. When they decided to listen to the tape after the session, they heard a voice.

> *"Hello?" a man's voice said. "Can you hear me? Can you hear me. This is Thomas calling. Can you hear me? Hello."*

The voice continued,

> *"I have for some time been involved with this experimental communication system, and it is this system that we are using now. It is hoped that in the future, this system will help us communicate with distant dimensions... and with all those we are calling upon will make themselves to be of service."*

The communication ended after approximately 20 minutes. At that point they realize that the T.A.E. could be Thomas' initials, Thomas Alva Edison.

On February 14th, the team received messages through the tape recorder without the Germanium device attached. Foy later suggested that their recordings were not unlike those documented by modern day ghost hunters. He wrote, "Perhaps there are 'doorways' involved here, if so, it would be a possible explanation for EVP (Electronic Voice Phenomenon) which many people report experiencing."

The SCOLE experiment was as controlled as possible by the mediums involved. It wasn't equivalent to a research experiment with double blind trials (where neither the researcher nor the participant knows what part of the trial they belong to). Other researchers have taken up that challenge and are testing mediums under more controlled conditions.

Gary Schwartz Research

Gary Schwartz is a psychology professor at the University of Arizona. He is particularly interested in mediumship and healing and is developing ways to scientifically test the reliability of mediumship. His first research investigations involved testing the veracity of mediumistic communication. His initial single-blind research was funded by the university and HBO, which filmed the testing. Five well-known mediums took part in the experiment and did readings for two people who recently had relatives who had passed on. The mediums were separated from the subjects so they would not receive clues from body language. The subjects were only allowed to answer yes or no to comments so as not to influence the mediums.

As with any good hypothesis test, there was a control group and a test group. The test group was composed of readings (messages from the spirits) by proclaimed mediums. The control group consisted of readings given by non-mediums. Correct interpretations were recorded and ranked. Results showed a 77% and 83% success rate for the mediums, and a 36% success rate by the non-mediums. In a second double-blind trial, a medium from Arizona gave a reading to a person in Los Angeles over the phone. Schwartz's testing was later published in *The Journal of Psychical Research* and his book, *The Afterlife Experiments* (2002).

All scientific research is subject to comments and critiques. Schwartz's first research was criticized for its flaws. He conducted additional research, including, "Anomalous information reception by research mediums demonstrated using a novel triple-blind protocol," which was published in *Explore* in 2007 with Julie Beischel. This was a triple-blind study designed to test mediums. Eight University of Arizona students served as test subjects Four had experienced the death of a parent, and four others the death of a peer. Eight mediums who had previously demonstrated an ability to report accurate information in a laboratory setting conducted the readings.

The methods of this study were more complicated in order to eliminate bias on the part of the researchers, the mediums, and the test subjects. The test subjects were not present at the readings; an experimenter was blind to the information about the subjects, and the deceased peer served as a proxy sitter. After the readings, the subjects read and rated the reading transcripts, one for them (intended) and the other the control. The findings included significantly higher success for "intended" versus "control" readings.

The results suggested that certain mediums can receive accurate information about deceased individuals. The study design effectively eliminated other forms of communication like body language or telepathy. It could not distinguish among alternative paranormal means of communication, such as survival of consciousness after death, super-psi or super-ESP, or retrieval of information via a psychic channel. It was a good start, but more tests need to be conducted.

Chapter 22
Reincarnation Investigations

That the soul or consciousness can survive death to return in another body is a common belief that has existed at least since the origin of Hinduism. Worldwide surveys show that most of world`s population believes in an afterlife, including 46% to 60% of those with no religious affiliation. The belief in reincarnation also exists in many religions and cultural systems. This includes the rebirth of a soul or spirit, the continuation of a stream of consciousness, or the migration from one form of being into another.

Because reincarnation memory recall is entirely subjective, the study of the topic involves the collection and comparison of stories after the fact. Methods include interviews with subjects, their current parents, relatives, and friends, as well as with alleged acquaintances of the subjects during their past life. Researchers use available documents, objects, photographs, and other records to corroborate the subject's stories. Some studies were also replicated to determine story consistency.

Past life research has found that children usually begin to make statements about a previous life when they are about 35 months of age. About 20% remember an interval between the two lives, 75% describe their mode of death, and 35% have phobias related to their mode of death. Others demonstrate skills they never learned, unusual behaviors, or birthmarks or birth defects related to their past life. Some remember their own name and names of people whom they knew, as well as places where they lived.

James Leininger

James Leininger's recollection represents a detailed account of a reincarnation. When he was 22 months old, James Leininger's father took him to the Cavanaugh Flight Museum near Dallas. The boy had no exposure to WW II planes and became fascinated with both the planes and a video they purchased, *It's a Kind of Magic*, about the Blue Angels. Within two months of the trip, James started slamming his toy planes into the coffee table and saying, "airplane crash on fire." At about the same time, he began having dreams of a plane crash. "Airplane crash! Plane on fire! Little man can't get out!"

James told his parents about his dreams. He said he remembered when his plane caught fire and crashed, that it had been shot by the Japanese. When he was 28 months old, he told his parents he had flown his plane, calling it a Corsair, off a boat. He said the boat was called Natoma. His parents thought the name remembrance was very specific for the young boy. His father searched online and found a description of the USS Natoma Bay, a carrier stationed in the Pacific during WWII. Intrigued, they asked him other questions about the ship and the "little man." They discovered that the name of the pilot in the crash was named James, just like their son. They also found that the pilot, James, had a friend, Jack Larsen, on the ship.

When James was two and a half years old, his father bought a book titled, *The Battle of Iwo Jima 1945*, as a Christmas gift for his own father. James pointed to a picture in the book and said that's where his plane crashed. Eventually, they found that Jack Larsen survived his tour of duty. Only one pilot had been lost during the war, a 21-year-old from

Pennsylvania named James M. Hudson, Jr. Jack Larsen was flying beside James when he was shot down.

James's nightmares continued. When he was old enough, he drew hundreds of battle scenes involving planes. He said that he signed "James 3" because he was the third James. Since James Hudson was a junior, that made James Leininger the third James.

James' story is an intriguing recollection, but it didn't happen in a controlled situation. Was there confirmation bias happening? Did the parents unconsciously only recall the memories that matched events supporting the story? Was James receiving unconscious cues from the people who knew more about the situation than he did? I'm not saying the memory is false. I'm saying that we can't be sure that it's true.

Ian Stevenson

Psychiatrist Ian Stevenson, from the University of Virginia, has investigated many reports of children who claimed to remember past lives. His interest in the topic all began in the 1920s, when he was growing up in Canada. He was a sickly child who suffered with bronchitis numerous times and spent hours in bed. During that time, he read his mother's collection of books on Eastern religions and was exposed to reports of paranormal phenomena. He eventually earned his medical degree at McGill University in 1943, before moving to Arizona where he briefly studied biochemistry. Then he studied psychosomatic medicine before training in psychiatry and psychoanalysis.

Stevenson's academic career flourished, and he became chairman of the Department of Psychiatry at the University of Virginia in 1957. Around that time, he also renewed his interest in the paranormal and wrote book reviews and articles for non-academic publications covering the topic. In 1958, he won the American Society for Psychical Research's contest for his essay, "The Evidence for Survival from Claimed Memories of Incarnations." The most convincing cases involved 44 young children, generally between the ages of 2 and 5 in Asian countries where the belief in reincarnation was accepted.

That original study led to many more. He collected data for more than 2500 case studies over a 40-year period and published a dozen books, including *Twenty Cases Suggestive of Reincarnation and Where Reincarnation and Biology Intersect.* Stevenson methodically documented each child's statements, identified their previous identity, and verified facts that corresponded to the child's memory. He also matched birthmarks and birth defects on children with wounds and scars on the deceased. These were verified with medical records such as autopsy photographs and published in *Reincarnation and Biology Vol. 1.*

Stevenson was diligent in his data collection and searched for disconfirming evidence and alternative explanations for the children's memories. When skeptics complained that most of his cases originated in Eastern societies, where the concept of reincarnation was accepted, he published a book, *European Cases of the Reincarnation Type.*

Stevenson was cautious about the data, stating that the evidence was "suggestive" of reincarnation, but not definitive. Despite that, he felt the information should be taken seriously.

Although a significant amount of data has been produced by past life memory studies like Dr. Stevenson's research, there remains much controversy about the explanations of these memories. Alternate explanations for memories include children's fantasies, fraud, socio-psychological needs of the children's families, or inherited memories as the cause of their responses. To move the research field forward, a way to evaluate past life memories must be established. This phenomenon is difficult to control because it occurs to individuals in normal household situations where the people involved rely on their own memories.

Jim B. Tucker

Psychiatrist Jim B. Tucker is director of the Division of Perceptual Studies, a section within the University of Virginia medical school's psychiatry department. Tucker, who was raised Southern Baptist, had no interest in reincarnation until he read Stevenson's, *Children Who Remember Previous Lives: A Question of Reincarnation.* Tucker was intrigued by the possibility of studying the children's reports in a systematic and objective fashion by verifying them through official records and witness interviews. He began working with Stevenson half-time for the Division of Perceptual Studies in 1999 as an assistant professor.

When Stevenson retired in 2002, Tucker assumed reincarnation studies, later becoming director. He and his research team continue investigate reports, but with an emphasis on American cases. At this point in time most of the 2500 cases have been entered into a database to make it easier to statistically analyze the accumulated data to find patterns and trends. While continuing Ian Stevenson's work, Tucker has published three books and numerous papers on the topic.

While Stevenson focused on reincarnation cases in Asia, Tucker studied U.S. children. Knowing that bias or the appearance of bias in such studies proves to be a problem, Tucker developed a Strength of Case Scale to evaluate reincarnation cases. The scale included four criteria: 1. Birthmarks or defects that correspond to the previous life injury, 2. Verification of statements about the past life, 3. Current behaviors related to the previous life, and 4. If the child could naturally be exposed to information about the past life.

In a research article published in 2000, Tucker evaluated 799 cases with his scale, testing two opposing hypotheses. One was that the reincarnation stories could be accounted for by natural causes. The second was that the stories were due to a paranormal explanation. He found that his scale was internally reliable, and more consistent with the paranormal explanation. He did conclude that "While they are not definitive, they do indicate that more research is justified in the cases of children who appear to remember previous lives."

Paul Edwards

In science, it is important to have sceptics to challenge hypotheses, methods, and conclusions. Skeptic, Paul Edwards, analyzed many of Stevenson's accounts and called them anecdotal. He and others suggest that selective thinking and false memories may account for the evidence and they cannot be taken as factual accounts. Others think the evidence may be cherry picked to create believable stories. Stevenson acknowledged these limitations but believed that a methodical approach was still needed to evaluate the phenomena.

Christopher C. French

Christopher French was born in England in 1956 and is a British psychologist specializing in the psychology of the paranormal. As the head of the University of London's Anomalist Psychology, he appears regularly in the media as an expert on testing paranormal claims. French considers himself a skeptic, despite conducting some of his own research on past-life memories. He began his career studying mainstream neuroscience before embracing anomalistic psychology, the study of human behavior associated with the paranormal. French assumes that nothing paranormal is involved.

French believes that most reincarnation reports can be explained as children experiencing false memories, though he respects Stevenson's meticulous research. He believes that young children will often say things that don't make sense, and the parents will then inadvertently feed them information trying to understand and that attempt creates false memories.

French believes research is crucial. It's important to determine if there is something genuinely paranormal happening or if this has something to do with normal human psychology. At the same time, mainstream science still relegates parapsychology to its own separate field, which prevents it from being evaluated with as much scrutiny as it should.

False Memories and Conformation Bias

Many psychologists have trouble believing reincarnation memories at face value because they know that memories are not entirely reliable, and some memories can be created. These are known as false memories. False memories differ from simple memory errors. Everyone experiences memory failures, especially about events that happened long ago. False memories are unique in that they are a distinct recollection of something that did not actually happen. Not only are they false, but they may also be very vivid, making the person very confident that the memory is real.

Psychologists have studied how false memories are created in laboratory situations where events are well controlled. These experiments give us insights into how false memories work. First of all, not all people are likely to form false memories. Children and older adults are more suggestible than people of college age in most false memory trials. Suggestibility also tends to increase with higher scores on the Dissociative Experiences Scale (DES), a measure of how distracted a person is. The most striking examples of false memories are caused by a combination of factors.

One classic case by Loftus et al. in 1995 involved implanting a false memory of having been lost in a mall as a child. The experimenters asked for the help of close family members, who told them several true events that each subject had experienced. When the subjects came into the lab, they were interviewed about the three true memories told to them by family members and a new false one about being lost in the mall introduced by the researchers. Subjects were interviewed several times over several months about the same memories, and by the end of the experiment more than a quarter of them remembered being lost in the mall. Loftus also discovered that memories become distorted and begin to change over time. In some cases, an original true memory may be changed to incorporate the false one.

Another problem with collecting reincarnation memories is that people are subject to confirmational bias without even knowing it.

So, what happens when a person or family or researcher is trying to determine if a child is having reincarnation memories? Are they taking every memory into consideration, or are they only looking at the memories that agree with the reincarnation story they are developing? Many reincarnation stories are convincing, but how many are the result of information that was given to the child and became a false memory? And how much information is being ignored by confirmational bias? Controlled research is really needed to eliminate these problems, and this is difficult to do when children begin to develop their reincarnation memories at such young ages in the home.

Chapter 23
OBE Observations

The OBE (Out of Body Experience) phenomenon challenges our conception of self and consciousness, which has led to the development of multiple theories and approaches to explain it. OBEs can be spontaneous but also induced through different methods. Those experiencing spontaneous OBEs tend to state that their experiences seemed more real than a dream, which is not always present in induced cases.

OBEs can be divided into two broad categories. The first posits that the experience involves an actual separation between consciousness and the physical body. From this perspective, consciousness would not be located in the brain but could exist independently. The second category interprets OBEs as internally generated mental phenomena such as dreams, imagination or hallucinations.

Michael B. Sabom

Michael B. Sabom was a cardiologist and an assistant professor of cardiology at Emory University before moving into private practice and becoming part of the staff at Northside and Saint Joseph's Hospitals in Atlanta, GA. It was during his medical residency in the 1970s that a social worker introduced him to a book about NDEs. He made it clear that he thought "the whole idea was absolute hogwash," but the social worker challenged him to investigate. None of the doctors knew what he was talking about, but when he began his investigation, the third patient he questioned told him about an NDE he had experienced.

Sabom focused his investigations on OBEs, assuming he could easily discredit the experiences. He spent the next five years collecting NDE accounts and comparing them to other studies. The NDEs were consistent in each study and corresponded to one or more elements in Moody's classic list reported in *Life after Life*. Sabom recorded three OBE patterns: autoscopic (a self-visualizing OBE), transcendental (entering a dark region, seeing a light, etc.), and combined (autoscopic followed by transcendental). Standardized NDE scales were used to assess the "depth" of the experience.

His first book, *Recollections of Death: A Medical Investigation*, published in 1981, was groundbreaking in the field of near-death research. The book was positively received for providing a relatively objective point of view and has been cited over 800 times in scholarly works.

In his first study, 32 subjects described their OBEs. Twenty-six gave broad visual impressions of their resuscitation, six claimed to have seen specific details of the resuscitation procedure such as oxygen mask placement, cardiac massage, defibrillator paddles, injections, checking for pupil responses, and drawing blood. When compared to the medical records the descriptions were found to be surprisingly accurate.

Sabom tried an experiment next. He hypothesized that if OBEs were imaginary, OBE patients' descriptions would be no more accurate than those described by nonexperiencing-OBE people who were asked to imagine what had happened during their cardiac arrest. Both groups of patients were asked to recall their own resuscitation as best they could. Sabom found that the group of OBE patients were much more accurate than the nonexperiencing-OBE control group in describing their own resuscitations.

Another OBE study with similar methodology to Sabom's study was published by Dr. Penny Sartori. This study also found that OBE patients were remarkably accurate in describing details of their own resuscitations. The non-OBE control group could only guess what occurred during their resuscitations.

Karlis Osis

Karlis Osis was born in Riga, Latvia in 1917. He is one of the first psychologists to obtain a doctorate in the field of ESP from the University of Munich in 1950. As a Research Associate of the Parapsychology Laboratory at Duke University, he was a colleague of Dr. J. B. Rhine in the 1950s. He later served as Director of Research at the Parapsychology Foundation and was the American Society for Psychical Research's Director. He authored numerous scientific articles on his research, which have included studies of ESP in animals, the effects of space and time on psi, ESP and meditation, physiological correlates, out-of-body experiences, and apparition phenomena. He then became the Society's senior researcher until his retirement in 1983 and died in 1997.

Otis approached the OBE question by using psychics, Alex Tanous and Ingo Swann, who claimed to be able to travel out-of-body and observe objects in distant locations. In one study he found that EEG amplitude was higher during their out-of-body periods. In another experiment, which took place in a perception laboratory, Tanous relaxed on cushions in a darkened room. An optical illusion device was positioned four rooms away. It contained a spinning disc divided into four quadrants, four colors, and five drawings. The device randomly stopped on quadrant, color, and symbol which could only be seen by an observer if they stood by the device and looked through a spy hole.

Tanous would have to identify what was being displayed. At the end of the trial, the device's selections would then be decoded to see whether Tanous's observations were correct above the level of chance. In addition, a strain-gauge was placed next to the optical device to see if it activated during the experiment. In 197 trials over twenty sessions, Tanous correctly identified the image 114 times and incorrectly 93 times. The number of identifications was found to be statistically significant. There was also strain-gauge activation during periods when Tanous produced a sequence of correct identifications. These experiments were a good start at trying to test if OBEs can be measured but had some flaws and would need to be repeated to be verified.

Chapter 24
Related Experiences

NDEs and OBEs are often explained away as visions or hallucinations caused by the dying brain. Since we don't understand how consciousness is linked to or generated by the physical brain, and we don't know exactly how the dying brain responds to the loss of oxygen and blood flow, it is entirely possible that NDE memories are created by some biological process. It's important to look at all aspects of the brain and its relationship to death and dying. In this chapter we will look at other experiences that occur and are not regularly discussed.

Bystander Experiences

Sally was with her mother when she died. The first thing Sally witnessed when her mother died was that the light intensity in the room changed. Many things happened at once. Her body rocked forward, and she was suddenly seeing the room from a different angle, above and to the left side of the bed. The event was disconcerting, but she felt comfortable and peaceful.

During the event, she wasn't sure if she was out of her body because other things attracted her attention. She watched scenes, like a 3D movie of her mother's life, flash throughout the room. The events seemed strange and perfectly natural at the same time. Some of the actions she remembered, others she didn't. She could see her mother watching the scenes too.

While the scenes were in motion, her mother left her body. Sally saw her father, who had passed away seven years before, helping the mother out of her body. He looked like a young man, not the 75 years of age that he was when he died. The room changed angles again, and everything went back to normal.

The above account is published in Raymond Moody's book, *Glimpses of Eternity*, in which he shares many stories of people who have had Shared Death Experiences (SDEs). Sally was a woman he met at a conference in New York. Like others in her situation, bystanders are neither ill nor injured. Unlike NDEs, SDEs are not caused by brain hypoxia or atypical seizures.

Bystanders may include physicians, nurses, medical personnel, relatives and friends of the dying. They often report SDEs that are indistinguishable from near-death experiences. Some see a transparent replica of the dying person leave their body at the time of death. Others describe leaving their own bodies and rising with the departed, part way toward a glowing light. Reports include brilliant lights filling the room, indescribably beautiful music and apparitions of the dead person's deceased loved ones. Like Sally, others report that they co-lived the witnessed a life review of the person.

Raymond Moody has interviewed hundreds of people who had bystander experiences and characterized them by the following:

1. Seeing the dying person's spirit leave the body as a golden-gray mist or transparent replica.

2. Accompanying the dying person's spirit part-way toward a light before returning to their own body.

3. Experiencing a change in geometry of the room and viewing it from an impossible angle.

4. Hearing music so beautiful that it is difficult to describe.
5. Seeing an unearthly light that fills the room with comfort.
6. Seeing spirits enter the room, apparently the dying person's deceased relatives.
7. Co-living the life review of a dying person that may take the form of a panorama.

Investigators are just beginning to understand dreams and visions that people experience during the death of a loved one. Research has shown that experiences such as End-of-life Dreams and Visions (ELDVs) may play an important role by offering comfort, psychological/spiritual solace, and meaning to the witness. The ELDVs involve mental and sensory activity while the patient is asleep (dreams) or awake (visions) and are typically reported to include seeing or feeling the presence of deceased loved ones. Research has evolved from case-based anecdotes and caregiver second-hand accounts to reports of patient perspectives. More recently, research expanded to explore the impact on post-traumatic stress and bereavement outcomes of familial caregivers.

In a 2020 paper, Rachel M. Depner published a qualitative analysis of 548 ELDV reports from 55 people enrolled in a hospice homecare program from January 2013 to March 2015. The study suggested that ELDV content may include a broader collection of experiences that reflect waking life than previously believed. She stressed the clinical implications of taking ELDVs seriously and listening to family members who experience such events. ELDVs were said to have the potential to provide comfort and closure. It is imperative that clinical staff be trained to discuss and validate ELDVs to reduce the stigma around this death experience.

In another study, the Shared Crossing Research Initiative analyzed written accounts and transcripts of interviews with 107 people reporting a total of 164 SDEs. They found that 4 distinct modes of an SDE occur: remotely sensing a death, witnessing unusual phenomena, feelings of accompanying the dying, and feelings of assisting the dying. These experiences led to 3 major effects: changes in belief, the reconciliation of grief, and the perception of continued relational bonds with the deceased

Investigations have shown that ELDVs are quite common and are beneficial experiences for the survivors. But are these ELDV events real or a hallucinatory way for the person to cope with the death of a loved one? ELDVs have often been dismissed by medical professionals as some sort of delirium. Delirium usually includes a breakdown in attention and awareness, but those experiencing ELDVs typically demonstrate organized thoughts.

End of Life Experiences

Near the end of her life, a hospice patient named Mary was found cradling an unseen baby in her arms. She called the baby Danny. He was a stillborn son she'd given birth to decades before. She had not spoken about him since the loss. Before her death, he came to her as a form of comfort. Dr. Christopher Kerr witnessed the event.

Christopher Kerr was born and raised in Toronto, Canada and has an undergraduate degree in Psychology, a Doctor of Medicine, a PhD in Neurobiology and completed his residency in Internal Medicine at the University of Rochester. Since then, he has become chief executive and medical officer at the Center for Hospice and Palliative Care in Buffalo, N.Y.

Kerr said in a TED talk that he is not a spiritual person. When he was 12 years old, his father insisted that they hurry to catch an airplane so they could go fishing. Kerr didn't see what the rush was about, but his father passed away soon after that at the age of 42. It was as if he had known of his impending death. When Kerr began working at a hospice in 1999, he discovered that many people had dreams and visions before dying. Since then, he has spent several years exploring what he terms end-of-life experiences (ELES) and how they impact both the dying and their loved ones. His goal is to help improve the patient's comfort at the end of life and the quality of hospice care.

Kerr has conducted 1400 interviews with dying patients and found that 80% of them have visions and dreams and 72% of those are with loved ones who have passed. These experiences bring comfort to the patient, and the frequency of occurrences increases over time until their death. He doesn't discuss the cause for these visions but sees them as a way to help patients transition to death.

Terminal Lucidity

Anna Katharina Ehmer, was born in 1895 in Germany. After suffering from meningitis as a baby, she never recovered. She didn't speak, gobbled down her food when it was given to her, and soiled herself. She was sent to a psychiatric institution called Hephata where she lived until her death in 1922. Anna would have been one of thousands disabled patients lost to the annals of time if she had not begun singing just before her passing. The doctors and hospital staff were rendered speechless when they heard her. One of her doctors, Friedrich Happich wrote:

> *One day I was called by one of our physicians, who is respected both as a scientist and a psychiatrist. He said: 'Come immediately to Käthe, she is dying!' When we entered the room together, we did not believe our eyes and ears. Käthe, who had never spoken a single word, being entirely mentally disabled from birth on, sang dying songs to herself. Specifically, she sang over and over again, 'Where does the soul find its home, its peace? Peace, peace, heavenly peace!' For half an hour she sang. Her face, up to then so stultified, was transfigured and spiritualized. Then, she quietly passed away. Like myself and the nurse who had cared for her, the physician had tears in his eyes.*
>
> *We witnessed the dying of this girl with deepest emotions. Her death posed many questions to us. Obviously, Käthe had only superficially participated in all that happened in her surroundings. In reality, she had apparently internalized much of it. Because, where did she know the text and the melody of this song from, if not from her surroundings? Moreover, she had comprehended the contents of this song and used it appropriately in the most critical hour of her life. This appeared like a miracle to us.*

For decades, researchers, hospice caregivers and surprised family members have witnessed people with Alzheimer's or other forms of dementia suddenly regain their memories and personalities just before death. It wasn't until 2008 that German biologist Michael Nahm at the institute for Frontier Areas of Psychology and Mental Health coined the term "terminal lucidity." Terminal Lucidity occurs when patients with dementia, strokes, brain tumors, or other psychiatric illnesses like schizophrenia, change their behavior prior to death. For example, people with dementia unexpectedly return to a clear mental state shortly before death. Those who have not been able to communicate suddenly show improvements. Others may hold prolonged eye contact or spontaneously talk.

Nahm has studied the phenomenon for years and has published about a dozen journal articles on the subject which examine case studies. For one article, he was able to retrieve 49 case reports, most recorded before 1849. He noted that few reports of terminal lucidity were published after that time and that recent publications focus on patients suffering from schizophrenia and dementia. He recommended more in-depth studies on psychopathology and neuropathology involved in terminal lucidity to improve therapies and a better understanding of the subject.

Hahm's 2012 review includes the case of a 5-year-old boy with a malignant brain tumor who spent weeks in a coma. The boy suddenly regained consciousness and said goodbye to his family before passing the next day. In another case, a woman with meningitis who had been incoherent suddenly "came to herself" and answered questions just before passing. An aggressive Alzheimer's patient who had lost so much of his memory that he didn't know his own name, became calm and spoke to a minister the day before he died.

Lim Chi-Yeon studied terminal lucidity at a teaching hospital. Of 338 deaths that occurred during the study period, terminal lucidity was identified in 6 cases. Lucidity lasted from several hours to 4 days and after experiencing terminal lucidity, half of the patients died within a week, and the remainder died within 9 days. He found that it is an unpredictable end-of-life experience, and that more attention should be directed toward understanding the condition to improve end-of-life care.

What is causing Terminal Lucidity? Some suggest that neurons may regrow in patients with damage, accounting for their lucidity. Although some regrowth is possible, it seems unlikely that the brain would be able to repair enough damage to the point that a person regains their past personality or memories. This also doesn't explain the short timeframe of terminal lucidity right before death.

Nahm hypothesizes that shortly before death, large changes in signaling among brain neurons might temporarily reactivate pathways that have been disrupted due to damage and disease. Maybe neurotransmitters might be produced in high amounts as some type of "pre-death" activity. This burst of neurotransmitters would be temporary and occur shortly before death. Of course, this hypothesis needs testing.

Is Terminal Lucidity caused by cellular or chemical changes in the brain, or something else? It definitely shows that we have a poor understanding of consciousness. We need more research to find out.

Chapter 25
OBEs, NDEs and the Brain

Most neurobiologists are materialists, working under the hypothesis that consciousness is a manifestation of the physical brain. If that is the case, then OBEs and NDEs are a product of the dying brain, not the result of consciousness separating from the body at the time of death. With modern resuscitation techniques and machinery to measure brain activity, we are getting a better understanding of what might be happening to the physical brain during and following death. What does that tell us about consciousness? Is it traveling out of the body or is the brain creating an imaginary world as the cells struggle to survive?

In the past, death was defined as the moment when the heart irreversibly stops beating. But now we have machines to keep the heart and lungs functioning. Since then, death has been defined as the cessation of brain activity. But is that truly the death of the body? Recent studies suggest that the brain can be flatlined (not producing brain waves) but still survive in animals for seconds to hours.

The Dying Brain

Jimo Borjigin was born and raised in Hohhot, the capital of the Inner Mongolia, part of the Autonomous Region of China. She received a BS in Physics and master's degree in Biophysics from Tohoku University in Sendai, Japan. In 1994, she received her PhD in Neuroscience from the Department of Neuroscience at Johns Hopkins University, followed by postdoctoral training in the laboratory of Dr. Solomon H. Snyder. In 2003, Dr. Borjigin joined the Department of Molecular & Integrative Physiology and now has a joint faculty position in the Department of Neurology at the University of Michigan.

Borjigin's team published ground-breaking work regarding brain activity of dying mammals and reported a surprising neurochemical surge that takes place during cardiac arrest. In their study with rats, they monitored brain activity with electroencephalography (EEG) while the rats were put into experimental cardiac arrest. The researchers measured several components of brain function and found that a temporary surge of brain waves (synchronous gamma oscillations) occurred within the first 30 seconds of the heart stopping. This activity exceeded levels found during the conscious waking state.

The team couldn't experiment on humans, but they reviewed the records of four patients who passed away from cardiac arrest while under EEG monitoring. All four of the patients were comatose and unresponsive. Their families decided to remove them from artificial life support. Like with the rats, there was a spike in brain activity after the machines were removed. The activity was also detected in the so-called hot zone of neural correlates of consciousness in the back of the brain. Does this account for the highly lucid mental experiences reported by human near-death survivors? We need more proof that the spike in activity is causing the OBEs and NDEs, but at least there is research questioning what happens in the dying brain.

Another scientist researching brain activity at the time of death is Ajmal Zemmar. When Zemmar was a young child, he and his family fled Afghanistan during the war with Russian and immigrated to Germany. He obtained a medical degree at the Goethe University-

Frankfurt am Main and a PhD in Neuroscience at the Brain Research Institute Zurich, jointly from the Swiss Federal Institute of Technology and the University of Zurich. His medical training took him around the globe from Germany to Switzerland, the United States, England, Canada, China and South Africa. Zemmar now specializes in epilepsy and functional neurosurgery, and his research examines the basis of death and near-death experiences and neuro-robotics.

Zemmar is a neurosurgeon at the University of Louisville. He knew that experimental animal studies showed increased brain activity after cardiac arrest. At that time, no study had investigated humans. He happened to have the records of an 87-year-old patient undergoing cardiac arrest after traumatic subdural hematoma (brain bleed) who was also being monitored with EEG. He found that after cardiac arrest, delta, beta, alpha and gamma brain waves were decreased but a higher relative percentage of gamma waves were observed. Although the brain injury had caused swelling, his data provided the first evidence from the dying human brain in a non-experimental setting. He said that gamma waves may signal that different brain regions are working together to produce sensations. How the brain does this, he says, "is one of the biggest mysteries in neuroscience."

There is the possibility that NDEs are some kind of unique hallucination produced by the last gasps of the dying brain, a burst of activity while the brain is in survival mode after losing oxygen. Another explanation for out-of-body experiences is that they might be caused by the vestibular system (provides balance and spatial orientation) of the brain.

OBEs and Medical Conditions

The out-of-body (OBE) experience is one of the hallmarks of the Near-Death Experience (NDE). But OBE-like experiences are not always associated with death and dying. Researchers have been studying the various causes for OBEs and the sensation of seeing the environment from outside one's body. OBEs have been reported in about 10% of the population in most of the world's cultures independent of a death experience and in association with medical conditions such as epilepsy, migraine and depersonalization. They are difficult to study because they usually occur spontaneously, are of short duration, and happen only once or twice in a lifetime. Investigations of neurological patients with OBEs are also rare, but there is some information available.

Orin Devinsky received his BS and MS from Yale University and MD from Harvard Medical School. He is a neurologist and Director of the NYU Comprehensive Epilepsy Center and Saint Barnabas Institute of Neurology and Neurosurgery. He is also a professor of Neurology, Neurosurgery and Psychiatry at NYU Langone School of Medicine. Devinsky specializes in the study of epilepsy and behavioral neurobiology and is active in the Epilepsy Foundation. He has dealt with traditional therapies, drug delivery devices and new medications to help those with the disease. In 1989, Devinsky and his associates described several patients with OBEs caused by brain damage and found that lesions predominantly in the temporal lobes (located behind the ears and responsible for processing sounds, memories and vision) were associated with OBEs.

Olaf Blanke is director of the Center for Neuroprosthetics and is the Bertarelli Foundation Chair in Cognitive Neuroprosthetics at the Ecole Polytechnique Fédérale de Lausanne (EPFL). He also directs the

Laboratory of Cognitive Neuroscience at EPFL and is Professor of Neurology at the University Hospital of Geneva. Blanke's research deals with the neuroscience of consciousness and embodiment, human augmentation, and personalized medicine. He also pioneered robotics and virtual reality technology in neuroscience of consciousness studies and cognitive psychology. His medical interests are dedicated to preventive, diagnostic and therapeutic procedures and devices for neurodegenerative disorders, mental health and well-being. Blanke is also co-founder of Metaphysiks Engineering and member of the board of Mindmaze.

Bianke and his team analyzed the brain lesions of several patients with OBEs and found that the temporo-parietal junction (TPJ) was affected in all their patients suffering with epilepsy and migraine. This is where the temporal and parietal lobes meet above and behind the ear. This area is critical for social cognition, attention and integrating information from different parts of the brain. On the basis of their findings, they suggested a model to explain OBEs, proposing that OBEs are related to a failure to integrate sensory information of one's body.

In another study, Lopez and Elziere traced OBEs to problems with the vestibular system, which consists of several parts of the inner ear. It is well known that problems with the vestibular system can cause dizziness, but in this study 14% of participants with dizziness from vestibular system disorders also had OBEs. Most of the patients experienced OBEs only after they started having dizziness for the first time.

The researchers concluded that problems with the vestibular system may lead to confusing signals being sent to the brain. OBEs are the brain's attempt to make sense of these signals. They believe that disturbances occur in the temporo-parietal region of the brain, including the TPJ. The TPJ plays a key role in self-perception and is responsible for integrating information from the external environment and from within the body. It can lead to illusions, including a sense of body reduplication, a difference in location between the body and the self, and a change in perspective to outside the body.

In two other studies, Jason Braithwaite and colleagues examined people with predisposition to OBEs in the normal population. These individuals were evaluated with the Cardiff Anomalous Perception Scale (CAPS) and the Launay-Slade Hallucination scale tests to measure proneness to hallucinations in university students. In the study, those with high CAPS scores, which are associated with temporal-lobe instability and body-distortion processing had more OBEs.

Descriptions of OBEs were associated with seizures in a study by Bruce Grayson and colleagues in 2014. The patients reported changes in emotional states (including sadness and apprehension), cognitive changes (including déjà vu, racing thoughts, confusion) other consciousness alterations (feeling tired, sleepy, "spacey," dazed, fatigued), sensory distortions (including flashing lights, wavy lines, insects, geometric colored shapes, and a variety of smells), and paresthesia (tingling or burning sensations). Other somatic sensations included headache, pounding in one's head, tightness in the head, head swimming, nausea, sweatiness, warmth, coldness, palpitations, feeling pulled or twisted, feeling one's energy drained, weakness, and stomachache.

Stimulating the OBE

Although medical explanations for OBEs are unlikely to be the only reasons why people experience OBE phenomenon, they indicate that malfunctions in the body's sensory systems can cause OBE-like symptoms. The study of the consciousness by neuroscience is in its infancy with no established models and very few data. Since OBEs are uncontrolled when they occur naturally, scientists are trying to stimulate the brains of patients to create OBEs to study.

Research by Blanke and associates have used transcranial magnetic stimulation on the temporoparietal junction (TPJ) of the brain. They found that magnetic stimulation can interfere with the ability of healthy individuals to imagine themselves in normal body orientation, like out-of-body experiences. Electrical stimulation of the TPJ in epileptic patients also produced various sensations associated with out-of-body experience.

Smith and Messier, in 2014, stimulated the brain's right angular gyrus (RAG) in a single patient who was undergoing evaluation for epilepsy treatment. Like TPJ, RAG also plays a role in integrating information, especially understanding physical relationships and is involved with attention, memory and language.

The patient was a 24-year-old woman who was a psychology graduate student at the time of testing. She reported that she had the ability to produce OBEs on demand and appeared surprised that not everyone had the experiences. She said she had the capability since she was a child. She discovered she could move above her body and used this as a distraction during the nap time at school. She continued to encourage this experience as an adult. It was often done before sleep as an aid to enter sleep. She was able to see herself rotating in the air above her body, lying flat, and rolling along with the horizontal plane. She reported watching herself move from above but remained aware of her unmoving "real" body.

She described the experience:

> *"I feel myself moving, or, more accurately, can make myself feel as if I am moving. I know perfectly well that I am not actually moving. There is no duality of body and mind when this happens, not really. In fact, I am hyper-sensitive to my body at that point, because I am concentrating so hard on the sensation of moving. I am the one moving – me – my body. For example, if I 'spin' for long enough, I get dizzy. I do not see myself above my body. Rather, my whole body has moved up. I feel it as being above where I know it actually is. I usually also picture myself as moving up in my mind's eye, but the mind is not substantive. It does not move unless the body does."*

This is the first study with a "normal" patient who could create an OBE-like experience on demand. There were limitations to the study. First, it involved only one subject, and second it relied on the participant's report of her own experience. They believed that the existence of such a case raises the possibility that this phenomenon may be more common than we think and is unreported because people do not think it is exceptional.

Drugs and OBEs

Researchers are also using drugs to evaluate OBEs and NDEs. One of those scientists is Steven Laureys. With over 25 years of experience, Laureys has studied neuroplasticity and the power of the mind using neuroimaging and neuromodulation techniques as well as hypnosis and meditation. He has served as head of the Centre Cerveau (Brain Clinic) at the Liège University Hospital in Belgium, and was founder of the

Coma Science Group, director of the GIGA Consciousness Research Unit at Liège University, and Professor at CERVO Brain Centre, Laval University, Canada. He also serves as Founding Co-Director of the Hangzhou International Consciousness Institute, China and as Chief Neurologist atTRAINM Neuro Rehabilitation Clinic, Antwerp. He has received awards from the European Academy of Neurology, European Academy of Sciences, German Max Planck Society and US Society of Cognitive Neuroscience. He is co-founder of Mind Care International Foundation.

With his team, he has published over 540 research articles on the workings of the brain. These include studies of the human mind in health and disease using the latest neurotechnology like MRIs and neuromodulation techniques like vagal nerve stimulation. He is renowned for his expertise in altered states of consciousness, cognitive disorders, neurorehabilitation, recovery of brain function, traumatic brain injury, concussion, coma, stroke, as well as near-death experiences, sleeping, memory and motor disorders, chronic pain and headache, and stress related symptoms.

In one of his experiments, he and his team wanted to determine if the drug N,N,-Dimethyltryptamine (DMT) could induce NDE-type experiences. They used thirteen healthy volunteers with the average age of about 34. They participated in a placebo-controlled, single blind study which consisted of screening and dosing for 2 sessions (different doses were being given to establish what was needed to induce a reaction), separated by 1 week. Participants reported feeling the effects of DMT soon after the injection. Results revealed that all 13 participants scored above the standard threshold on the Greyson Scale test for an NDE in relation to their DMT experiences. Fifteen of the 16 NDE items were rated significantly higher under DMT compared to placebo, with 10 of these reaching statistical significance after multiple testing correction.

Laureys stated that there were many limitations to the study. The participants underwent the placebo trial first and were familiar with the Greyson Scale test questions before their drug trial. That may have primed participants to experience NDE-like phenomena ahead of their DMT sessions. The positive scores on the NDE scale might also reflect a general tendency to endorse anything unusual in relation to the psychedelic experience. Even so, the strong overlap on the items of the NDE scale suggested that more testing needs to be done. They felt that better understanding of the dying brain by using psychedelics "may have implications for how we view this most inevitable and universal phenomenon, potentially promoting a greater familiarity with and healthy acceptance of it."

Ronald Griffiths was another premier researcher in the area of drug induced effects on consciousness and how they compared to NDEs. Griffiths was a professor of behavioral science and psychiatry who worked with psychedelics. As a distinguished psychopharmacologist and professor at the Johns Hopkins School of Medicine in Baltimore, he spent decades studying the mechanisms of dependence on mood-altering drugs. He published scores of papers on opiates, cocaine, sedatives, alcohol, nicotine and caffeine.

When his research, "Psilocybin Can Occasion Mystical-Type Experiences Having Substantial and Sustained Personal Meaning and Spiritual Significance," appeared in *Psychopharmacology*, it was a media event. *The Economist* called it "The God Pill." His research involved administering high doses of psilocybin to healthy people in a controlled setting. They described "profound feelings of joy, love and, yes, terror, along with a sense of interconnectedness and even an understanding of a sublime, sacred and ultimate reality."

Griffiths' work was endorsed by the Food and Drug Administration and the Drug Enforcement Administration and other experts. In 1999, he began to examine the nature of consciousness and religious experience. He compared psychedelic drug-induced and non-drug experiences of 3192 people, 933 in the non-drug group, 904 on LSD, 766 on psilocybin, 282 on ayahuasca, and 307 on DMT. Compared to the psychedelic groups, the NDE participants were unconscious, clinically dead, or their life was in imminent danger. He found that NDE and drug induced people were similar in their changes in attitudes toward death. There was a reduction of fear of death, an increase in positive feelings and personal meanings, spiritual significance, and psychological insight. The NDE participants were more likely to rate their experiences as the single most meaningful of their lives.

A paper published in 2019 by Charlotte Martial and others of the Coma Science Group at University Hospital in Leige, Belgium in Consciousness and Cognition, reviewed the similarity of about 15,000 scientific studies linked to 165 psychoactive substances and 625 NDE narratives. They determined that ketamine induced experiences were most like NDEs. They did state that their conclusions were based on personal reports and that "no physical examinations of the subjects at the time of the reported NDEs were performed." They also noted that more empirical evidence is needed to determine how the drugs produce NDE-like effects.

In the end, they concluded that "With the available experimental data, it is neither possible to corroborate nor refute the hypothesis that the release of an endogenous ketamine-like neuroprotective agent underlies the NDE phenomenology." They believe that ketamine could be a good drug to test patients for NDE responses in the future.

Since we do find OBE-like and NDE-like experiences in patients with medical conditions or under the influence of psychedelics, does that mean that we have found the explanation for OBEs in dying patients? Patients with seizures, epilepsy as well as those undergoing brain stimulation do describe a feeling of being out of their body, but it is unclear if they have other experiences like dying patients while in that state.

Dying patients experiencing OBEs during NDEs report events or objects that they could not have witnessed from their vantage point. These include details of their medical procedures or hospital rooms or staff while they were clinically unconscious or incapable of perceiving their surroundings.

In science, being skeptical is a good thing. Since OBEs and NDEs are personal accounts, sometimes being reported months or years after the event, many question the reliability of the accounts. No person's memory is perfect, and memories may change over time. Questions that need answered are: Is the event the result of post-hoc rationalization (making the story fit our expectations)? Is it the result of selective memory (only remembering the parts of the event that fit the experience)? Is there a subconscious awareness of details at play (not realizing the memories came from a different source)?

The Meditating Brain

William Van Gordon from the Centre for Psychological Research, University of Derby in the UK investigated meditation-induced NDE-like experiences that are referred to in ancient Buddhist texts. These ancient texts describe meditation practices intended to help spiritual practitioners prepare for the processes of dying and death. They include the eighth-century *Tibetan Book of the Dead* and *Profound Dharma of Natural Liberation Through Contemplating the Peaceful and Wrathful: Stage of Completion Instructions on the Six Bardos* by Padmasambhava. The books outline specific phases that consciousness progresses through during and following death. Another text used was, *Delok: Journey to the Realms Beyond Death* by Drolma, that recounts the author's experience of being a *delog* (individuals that experience NDEs and share them with the masses). The Dalai Lama has also asserted that experienced meditators could induce this state so when it naturally occurs at death, they can recognize and sustain it.

Van Gordon suggested that since some advanced Buddhist meditators can induce NDEs, that this may be a way to study them more scientifically to understand the psychology and physiology of death-related processes. His research in 2018 studied 12 advanced Buddhist meditators for three years and compared their Meditation-Induced NDEs (MI-NDE) against two other meditation practices in the same participant group. Changes in the content and insight after the NDEs were assessed over a 3-year period.

They found that those practicing MI-NDEs increased their insight, mystical experiences, and non-attachment. The participants were 1) consciously aware of experiencing MI-NDEs, 2) retained control over the content and duration of MI-NDEs and 3) elicited a rich array of non-worldly encounters and spiritual experiences. Comparing their results to the Grayson Scale, they found that participants experienced a genuine form of NDE and that outcomes relating to the MI-NDE condition were not simply the product of a standard or profound form of meditation.

The study did have limitations. The NDE scale relies on self-reported events and feelings. The sample was exclusively composed of Buddhist meditation practitioners, suggesting that interpretations of the MI-NDE due to religious predispositions were not controlled for. The researchers suggested that it would be ethically feasible for future research to recruit advanced meditators to assess real-time changes in neurological activity during MI-NDEs.

This chapter has looked at a variety of experiences that are associated with dying and death as well as the effects of drugs and meditation. They might suggest that changes in brain chemistry are responsible for NDEs and OBEs, but chemical changes have not been measured. So far, rather than converging into a unified theory, investigations into OBEs have led to several disconnected hypotheses. These include multiple explanations from reality distortion to brain pathology.

Vernon Neppe, Director of the Pacific Neuropsychiatric Institute in Seattle, WA, evaluated numerous OBE studies. He concluded that studies were limited because they hypothesize one single cause and outcome, whereas out-of-body experiences may have multiple causes and manifestations, rather than a single generalized explanation. Despite the increase in research on out-of-body experiences, understanding of the phenomenon is limited, both among the public and health professionals. Much more research is needed to evaluate both experiences and brain changes associated with OBEs.

SECTION VIII
THE HYPOTHESIS

Chapter 26
Is Testing Possible?

If we are going to develop a hypothesis that consciousness survives the death of the brain and moves into another state of awareness, it would be nice to define and identify consciousness. But it looks like we won't have that answer for some time. If we don't have a good measure for consciousness, then how do we determine if the consciousness survives death? There are two events that occur during NDEs. The first is the out-of-body experience (OBE) when the person rises above their body and watches what is happening. The second is the transition into another conscious state we can refer to as heaven or the afterworld. To determine if these events are real, we are going to have to use subjective descriptions and anecdotal stories, which are not typically used in scientific research because they may not be reliable. Is there a way to create a test that can confirm the subjective descriptions given by those experiencing the out-of-body part of the NDE?

Quantifying Out of Body Experiences

Many people who have returned from the brink of death say they have viewed their body from above, either at a crash site or during an emergency medical situation. Is the OBE experience some sort of hallucination or is it really occurring? Is it trauma to part of the brain?

Some anecdotal experiences are very convincing, but let's examine the topic objectively.

Here's what we do know. OBEs accompany near-death experiences 76 percent of the time. Are OBEs just spiritual experiences created by people's religious background? Some do attach spiritual connotations to the event, but many subjects regard their out-of-body experiences as a curiosity. They may call it odd or intoxicating, but not profound. Others are leery of their experience. As philosopher Descartes said: "Whatever I have up till now accepted as most true I have acquired either from the senses or through the senses. But from time to time, I have found that the senses deceive, and it is prudent never to trust completely those who have deceived us even once."

To validate the OBE as an actual occurrence, can testing measures be prepared to gather the highest quality data from a subjective event? The 1990 movie, *Flatliners*, took this question to the extreme. Medical student Nelson Wright convinces his classmates to help him discover what lies beyond death. In a clandestine experiment, Nelson is given medication to stop his heart and he flatlines for one minute before his classmates resuscitate him. Others follow Nelson's lead, extending their times of death. All experience visions, and these visions have dire consequences that haunt them.

No researcher would purposefully kill a patient to try to elicit an NDE or OBE, but there is a way to take advantage of ongoing emergency situations. When a patient is in the hospital or is transported to the hospital in a near death condition, machines that monitor heartbeat, brain waves, oxygen saturation of the blood, and other metabolic functions could be made available to gather data. We know that there can be a spike in brain activity just before death. An EEG can measure brain waves to determine what is happening at what time during the emergency event. Did the heart stop and the brain struggle to survive, creating a hallucination? Or was the brain completely silent at the time of the OBE?

The survivor's testimony is the most important evidence in corroborating an OBE. Unfortunately, accounts of OBEs are often given long after they happen, allowing the memories to change and fade. To ensure the most accurate recollection, interviews must be taken as soon after the event as possible. Rising out of one's body and witnessing the emergency team performing CPR is a general occurrence that everyone has probably seen on television shows multiple times. This may be what the person saw, but it is not specific enough to be sure the vision was not generated by the person's own memories and mind. There needs to be a way to determine if the person's consciousness was actually out of their body, like presenting them with a specific object or sign the person can recognize in the room.

To verify an OBE event, testimonies should be taken from others witnessing the emergency event as well. If the patient heard a conversation that was specific in the emergency room, observed a happening or conversation outside of the room, or saw something strange like a shoe on a ledge outside the hospital window, it needs to be recorded. Testimony from those observed, taken independently without them knowing the original story, is also necessary. Video cameras could even be used to validate events and conversations.

Researchers, like Raymond Moody, have compiled lists of NDE data to show the similarities of experiences between people across the globe. Now there needs to be a comprehensive way to determine if OBEs can be documented and verified. Let's look at one case to examine the promise and problems with newer medical techniques.

Sam Parnia and AWARE

Pam Reynold's OBE case is an example of why more rigorous studies are needed. That is why Dr. Sam Parnia has taken up the gauntlet. Sam Parnia is a graduate of Guy's and St. Thomas Medical School in London. After continuing his studies at the University of Southampton, he worked as a clinical research fellow and received a PhD in cell biology in 2007. Parnia joined the faculty at Stony Brook University School of Medicine as a member of their Pulmonary, Critical Care, and Sleep Division where he leads research on cardiopulmonary. He also founded and directs the Human Consciousness Project.

While at University of Southampton, Parnia was the principal investigator of the AWARE Study, which ran from 2008 to 2012. The study included 33 investigators from 15 medical centers across three countries. The purpose was to record consciousness, memories, and awareness during cardiac arrest when no blood was flowing to the brain.

To confirm OBE recollections, Parnia did more than interview the patients. He set up a testable situation to determine if patients were really out of their body during an OBE. It was quite simple. In the rooms where cardiac arrests were most likely to occur (50 to 100 areas of the hospitals), shelves were installed where cardiac resuscitation was most likely to occur. Each shelf held one image that included national or religious symbols, people, animals, and major newspaper headlines. The images faced the ceiling so that they were not visible to the patients or hospital staff. If a patient's consciousness left their body and observed the cardiac resuscitation from above, the images could be seen. They also installed a triangle on the underside of the shelf to determine if the patients thought they were seeing their bodies from above but were actually looking up from the bed at some time during their recovery.

Once the study began, Parnia soon understood the difficulty of collecting information. Out of the 2060 cardiac arrest events that occurred, 1730 of the patients did not survive. Of the survivors, only 140 completed stage 1 interviews while still an inpatient, and only 101 of those 140 patients completed stage 2 interviews. Parnia stated that part of the problem is that drugs given during and after cardiac arrest often suppress memory, and patients are kept sedated or in a coma after the event. Out of the 101 patients, 46% had memories with 7 major themes: fear, animals/plants, bright light, violence/persecution, Deja-vu, family, and recalling events post-cardiac arrest.

Only 9% of patients had NDEs, and only 2% (2) described an OBE, seeing and hearing events related to their resuscitation. One of the two OBE patients was too sick for the accuracy of her recount to be verified. It was possible to verify the accuracy of the experience of the second patient. Unfortunately, these two patients' cardiac arrests did not occur in areas equipped with test shelves, so no images could be used to objectively confirm their claims.

The results of the AWARE I study were published in October 2014. This led to the beginning of the next study, AWARE II. AWARE II is examining the relationship between the "quality of brain resuscitation and a patient's neurological, functional, cognitive, and consciousness outcomes following cardiac arrest." This time, the Parnia lab partnered with 20+ hospitals across the United States and United Kingdom to analyze patients' brain processes during cardiac arrest and resuscitation to "determine how consciousness may relate to brain function and survival, neurological, neurocognitive, and functional outcomes, and what it informs us about the experience of death."

They are also using new technologies. When a cardiac arrest occurs, they have a kit with portable machines, an oximeter that can measure the amount of oxygen in the brain and an EEG that measures electrical brain activity. They also have headphones that control what the patient is hearing and a computer tablet that generates visual images to determine if the patient is sensing visible cues. Like in the previous study, survivors undergo interviews to determine if they can recall any experiences.

In the study, published in October 2023, they wrote that 53 patients survived out of 567 cardiac arrest episodes. Of those who survived, 28 completed interviews, and 11 of those reported memories or perceptions. Their perceptions were classified as: emergence from coma during CPR or in the post-resuscitation period, dream-like experiences, and transcendent recalled experience of death. Nobody identified the visual image, and only 1/28 identified the auditory stimulus.

They noted that despite the loss of oxygen in the brain, normal EEG activity lasted as long as 35-60 minutes into the resuscitation event. This showed that conscious awareness could occur during cardiac arrest and may be the reason for near-death experiences. It may be that the dying brain loses inhibitions while trying to survive and creates the NDE. They suggested that NDE recollections and brainwave changes may be signs of "universal, shared elements of so-called near-death experiences."

Is that the final decision on the matter? The dying brain is creating hallucinations. Maybe not.

Dr. Bruce Greyson, Carlson Professor Emeritus of Psychiatry and Neurobehavioral Sciences at the University of Virginia School of Medicine in Charlottesville, brought up an important point. Two of the

28 patients interviewed had EEG recorded brain activity during their resuscitation, but they weren't the same patients who had NDE type experiences. Parnia's team did not show any association between active brain waves and conscious activity.

Parnia admits that patients who had NDEs did not show brain wave activity, and those who did show brain activity did not report NDEs. The main problem with the study is that most cardiac patients don't survive, so researchers don't have hundreds of survivors to interview. With over 500 cardiac patients, only about 10% survived and of those, only 2 remembered OBEs. Of the patients who did live and had readable electroencephalograms, 40% had brain waves that went from flatline to showing normal signs of brain function. Parnia concluded that "they have neither proved nor disproved that the experiences and claims of awareness during cardiac arrest are definitive." He plans to conduct more studies.

Chapter 27

Where are we now?

About fifty thousand years ago, in a cave located east of the Mediterranean Sea, a grieving mother and her small band of Neanderthals placed the body of her 18-month-old child in a shallow nook. It was laid on its right side, curled in a fetal position. Before the body was covered, the jawbone of a red deer was placed in the grave with the child. We may never know what the jawbone symbolized, but we do know that something special was taking place.

The area where the Neanderthal family buried their child was known as the Levant, now present-day Syria, Lebanon, Isreal, Palestine, and Jordan. In ancient times, it was an important migration crossroad linking Africa, Europe and Asia for both animals and early humans. Sixty thousand years ago, growing ice sheets in Europe turned the Neanderthals temperate climate into an arctic tundra. They were forced to move south and east into the Levant area where greater rainfall produced lush vegetation and attracted more animals to hunt. With more food resources the population grew, and so did their burial sites.

In a comparative study, Professor Been and Dr. Barzilai analyzed 17 Neanderthal burials in the Levant, including famous sites like Teshik Tash, Shanidar, Dederiyeh, Amud, Tabun, and Kebara. In some cases, large rocks were positioned around the body, maybe as a symbol of the boundary between the living and the dead. Some burials included modified limestone pieces placed under the head like a pillow. Others included grave goods such as goat horns, deer antlers, mandibles and maxilla. They may have been the first to believe in an afterlife, but they wouldn't be the last.

Fifty thousand years after Neanderthals buried their family members in the Levant caves, Predynastic Egypt began to develop. Early burials were in simple, shallow pits with a few burial goods, not unlike Neanderthals. Over time, graves became more complex, and bodies were placed in wooden or terracotta coffins. By 3500 BCE, mummification began, and rituals were performed to help the deceased enter the afterlife. If any culture was known for their dramatic mourning of the dead and burial rites, it was certainly Egypt. The Egyptians also had a written language, so we know exactly what they were thinking. After being judged by 42 divine beings for any wrongdoing during their lifetime, the deceased's heart was weighed to determine their virtue. If they passed, they could enter the Field of Rushes which was much like the real world they had just left with blue skies, rivers and boats for travel, gods and goddesses to worship and fields and crops that needed tended.

By the time that Galileo developed the idea of testing hypotheses in the 17th century, Christianity in the form of Catholicism, had spread across Europe. Beliefs in both heaven and hell were firmly established. As a matter of fact, religious dogma and control were so prevalent that new scientific discoveries which opposed church beliefs were considered sacrilege. After Galileo studied motion, he delved into astronomy. He supported Copernicus' view that the sun was the center of the solar system, which got him reported to the Roman inquisition. The Bible was clear; the sun revolved around the Earth. Galileo was found guilty of heresy by the Catholic church in 1633 and sentenced to house arrest for the rest of his life. Of course, we know now that he and Copernicus were correct.

Four hundred years later, we live with scientific advancements that would have astounded Galileo. When James Webb Space Telescope launched in December of 2021, it was the largest telescope in space. Equipped with near-infrared sensors, it can detect objects 100 times fainter than its predecessor, the Hubble Telescope. As it collects light from distant galaxies, it is not only seeing objects farther away but also objects farther back in time. To date, a galaxy named MoM-z14 has taken the title of most distant galaxy the telescope has found. It is a mind-boggling 13.5 billion light years away. It predates not only our existence, but the existence of earth.

In 1975, Raymond A. Moody, published his seminal book, *Life After Life*. Even though near-death experiences had been recorded as early as 2500 years ago, Moody was the first to bring the subject to light for the public at large. Since then, public interest has grown. Goodreads has almost 150 books listed under the topic of Near-Death Experiences. Some are collections of NDEs. Some are evaluations of the experiences, like books by Raymond Moody and Bruce Greyson. Others are personal stories of people's experiences. Some even claim they have *Proof of Heaven*, like Eben Alexander, MD. Unfortunately, most books are based on personal experiences, not on tested hypotheses.

Many people believe in heaven as an act of faith. That is what most religious tenets require. No proof is needed. For those who would like a little more convincing, personal reports of OBEs and NDEs may be enough to convince them that heaven exists. Scientists, on the other hand, are more like Thomas in the Gospel of John 20:24-19 (King James Version).

> *But Thomas, one of the twelve, called Didymus, was not with them when Jesus came. The other disciples therefore said unto him, We have seen the Lord. But he said unto them, Except I shall see in his hands the print of the nails, and put my finger into the print of the nails, and thrust my hand into his side, I will not believe. And after eight days again his disciples were within, and Thomas with them: then came Jesus, the doors being shut, and stood in the midst, and said, Peace be unto you.*
>
> *Then saith he to Thomas, Reach hither thy finger, and behold my hands; and reach hither thy hand, and thrust it into my side: and be not faithless, but believing. And Thomas answered and said unto him, My Lord and my God. Jesus saith unto him, Thomas, because thou hast seen me, thou hast believed: blessed are they that have not seen, and yet have believed.*

Jesus blesses those who "have not seen, and yet believed," but Thomas was the ultimate skeptic. And in this case, I mean a skeptic after the Greek, *skeptikos,* which means thoughtful consideration. He questioned the statements of others but was convinced with evidence. Unfortunately, today, we have many skeptics who cannot be convinced with any evidence. That's why science is structured with hypotheses that can be tested on rigorous, controlled ways, and statistical analysis that can limit bias.

Humankind has come a long way. From upright Homo erectus explorers, who adventured from Africa into Europe and Asia, to modern Homo sapiens who began making stone tools and reached the surface of the moon. The science that enables the James Webb Space Telescope to see farther into space than man has ever looked before, is the same science that can now be used to test for the survival of consciousness after death.

Of course, science can often take a long time to validate a hypothesis. Albert Einstein developed his famous General Relativity Theory in 1915. His predictions (hypotheses) of his theory have been tested several times with more and more sensitive experiments, just to make sure it is correct. As recently as 2016, the Advanced LIGO team directly detected gravitational waves coming from a black hole in February 2016. That is over 100 years after his theory was published.

The existence of the Higgs boson, an elementary particle of the Higgs field that gives other particles their mass, was proposed by physicist Peter Higgs in 1964. Could he prove the Higgs field existed? No, but the prediction looked good. Unfortunately, to find the Higgs particle, researchers needed a particle accelerator that was powerful and sensitive enough to detect it. It wasn't until the CERN Large Hadron Collider started up in September of 2008 in Geneva, Switzerland, that they had a chance of finding the Higgs boson. And even with the new machine, the particle wasn't confirmed until 2012. That was 48 years after it was proposed.

In the previous chapter, we looked at Sam Parnia's research on cardiac arrest survivors. In the latest experiment, it took 500 cardiac arrests to yield two OBE/NDE events. And those results were not definitive. It could be that proving the existence of heaven is much like finding the Higgs boson. It may take decades, or even a century, to have enough data to confirm or deny that consciousness exists independent of the body.

In Cueva de La Pasiegai near Puente Viesgo, Spain is a cave used first by Neanderthals and then Homo sapiens. One of the Neanderthal paintings is a red ocher ladder composed of red horizontal and vertical lines which dates back 64,000 years. One can imagine the artist creating a message in torch light, shadows wavering across the cold cavern walls. Will we ever know its meaning? Maybe not. But every time I see it, I think of the song, "Stairway to Heaven," by Led Zeppelin. Maybe the artist's message has traversed time after all. We have been searching for heaven for thousands of years. Maybe now we have the tools to find it.

Sources

Chapter 1 What is a Hypothesis

Carroll, S. (2019) *Something Deeply Hidden.* Penguin Random House, New York.

Creath, K. and G.E. Schwartz (2004) Measuring effects of music, noise, and healing energy using seed germination bioassay. *The Journal of Alternative and Complementary Medicine,* 10:113-122.

Dreger, A. (2016) *Galileo's Middle Finger: Heretics, Activists, and One Scholar's Search for Justice.* Penguin Books, New York; Reprint edition.

Freedman, S.J. and J. F. Clauser (1972) Experimental Test of Local Hidden-Variable Theories. *Physical Review Letters,* 28:938.

Greene, B. (2011) *The Hidden Reality.* Vintage Books Random House, New York.

Hassanien, R.H.E., et al. (2014) Advances in effects of sound waves on plants. *Journal of Integrative Agriculture,* 13(2):335-348.

Kaku, M. (1995) *Hyperspace.* First Anchor Books, New York.

Kong, J. et al. (2020) Measurement-induced, spatially-extended entanglement in a hot, strongly-interacting atomic system. *Nature Communications,* 11, Article number:2415

Lin, Y., et al. (2020) Quantum entanglement between an atom and a molecule. *Nature, 581 (7808):273.*

Parsonnet, J. (2005) Clinician-Discoverers — Marshall, Warren, and H. pylori. *New England Journal of Medicine,* 353:2421-2423.

Settle, T. B., *(1983) Galileo and Early Experimentation, Springs of Scientific Creativity, (F. Aris, H.T. Davis, and R.H. Stuewer, eds.).* University of Minnesota Press, Minneapolis. pp. 3-20.

Wang, B. C., et al. (2003) Biological effect of sound field stimulation on paddy rice seeds. *Colloids and Surfaces* (B. Biointerfaces), 32:29-34.

Chapter 2 Critical Thinking

Epstein, S. M. (1987) "Scholars Will Call it Nonsense" The Structure of Erich von Daniken's Argument. *Expedition Magazine,* 29(2).

Kirkland, S. D. (2013) *The Ontology of Socratic Questioning in Plato's Early Dialogues.* State University of New York Press.

Von Daniken, E. (1999) *Chariots of the Gods.* Reprinted Berkley Books, New York.

Wang, Y., et al. (2019). Systematic Literature Review on the spread of health-related misinformation on social media. *Social Science & Medicine.* 240: 112552.

Poland G. A. and R. M. Jacobson (2011). The age-old struggle against the antivaccinationists. *The New England Journal of Medicine.* 364 (2): 97–99.

Rosen, J. (2025) New Data shows MMR vaccination rate decline across the U. S. *Johns Hopkins Magazine,* June 2025.

Gordon, A., et al. (2026) Developmental convergence and divergence in human stem cell models of autism, *Nature* (29 January 2026).

Scriven, M. and R. Paul (1987). "Critical Thinking as Defined by the National Council for Excellence in Critical Thinking" The Foundation for Critical Thinking. www.criticalthinking.org.

Chapter 3 Let's Define Heaven

Brodie, F. M. (1995) No Man Knows My History: The Life of Joseph Smith. Vintage Press, New York.

Caluori, D. (2015) *Plotinus on the Soul,* Cambridge University Press.

Freud, S. (1949) *The Ego and the Id.* The Hogarth Press Ltd, UK.

Greene, B. (2024) *The Elegant Universe: Superstrings, Hidden Dimensions, and the Quest for the Ultimate Theory,* 25th Anniversary Edition. W.W. Norton & Company, New York.

Hanson, R. and R. Mendus (2025) *Buddha's Brain: The Practical Neuroscience of Happiness, Love, and Wisdom.* New Harbinger Publications, Oakland, CA.

King James Bible Standard Edition

LeDoux, J. E. (2023) *The Four Realms of Existence: A New Theory of Being Human. Belknap Press,* Harvard University, Cambridge, MA.

Lommel, Pim van, et al. (2001) Near-death experience in survivors of cardiac arrest: a prospective study in the Netherlands. *The Lancet,* 358, December 15, 2001.

Miller, L. (2010) *Heaven: Our Enduring Fascination with the Afterlife.* HarperCollins Publishers, New York.

Moody, R. (2015) *Life After Life: The Bestselling Original Investigation That Revealed "Near-Death Experiences."* Harper One, New York.

Pinch, G. (2004) *Egyptian Mythology: A Guide to the Gods, Goddesses, and Traditions of Ancient Egypt.* Oxford University Press, UK.

Quran, Amatul Rahman Omar, Al-Baqarah-The Cow-2:82 (Sura: 2, Verse: 82)

Randall, L. (2012) *Knocking on Heaven's Door: How Physics and Scientific Thinking Illuminate the Universe and the Modern World.* Harvard University Ecco, Cambridge, MA.

Remes, P. (2007) *Plotinus on Self. The Philosophy of the 'We',* Cambridge University Press.

Stead, W. T. and E. Stead (2013) *The Blue Island: and Other Spiritualist Writings (Life on Other Worlds)* Square Circles Publishing.

Swaminathan, G. (2020) *From the Beginning of Time: Modern Science and the Puranic Universe.* Notion Press, New York.

Trent-Von Haesler, N. and M. Beauregard (2013) Near-death experiences in cardiac arrest: implications for the concept of non-local mind. *Archives of Clinical Psychiatry* (Sao Paulo), 40:197-202.

No author (1869) "Location of Heaven" in *Scientific American* 21(11):170.

Chapter 4 Ancient Spiritual Beliefs

Bonato, B., et al. (2021) Cracking the code: a comparative approach to plant communication. *Communicative and Integrative Biology,* 14(1):176-185.

Eichler, A. (2011) Animism is actually pretty reasonable. *The Atlantic,* 24 January 2011.

Elkins, A.P. (1954) *The Australian Aborigines: how to understand them.* Sydney: Angus and Robertson.

Harvey, G. (2017) *Animism: Respecting the living world.* Hurst & Company.

Hodge, A. (2012) Plant root interactions. In *Biocommunication of Plants.* Springer, Berlin: pp. 157-169.

Khait, I. et al. (2023) Sounds emitted by plants under stress are airborne and informative. *Cell,* 186(7): 1328-1336.

Iupona, J. K. (2014) *African Religions: A Very Short Introduction.* Oxford University Press, UK.

McIntosh, M. A. (2019) A history of animism and its contemporary examples. *Brewminate,* 21 March 2019. Available at: https://brewminate.com/a-history-of-animism-and-its-contemporary-examples/

Perkins, M. K. (2019) What is animism? *Learn Religions,* 5 April 2019. Available at: https://www.learnreligions.com/what-is-animism-4588366

Tylor, E. B. (1891) *Primitive Culture: Researches into the Development of Mythology, Philosophy, Religion, art and Custom.* (2 volumes) John Murray, London.

Underhill, M. (2010) *The Hindu religious year.* Nabu Press, Charleston, SC.

"The Berber Culture" in *Moroccan Journeys.* Available at: https://moroccanjourneys.com/the-berber-culture/

SikhiWiki. (2020) "Sikhism and the environment" in Sikh Dharma International, 21 May 2020. Available at: https://www.sikhdharma.org/sikhism-and-the-environment/

Chapter 5 First Words about the Afterlife

Bottero, J. (2001) *Religion in Ancient Mesopotamia.* Trans. by Teresa Lavender Fagan. University of Chicago Press, Chicago.

Chavalas, M. W. (2003) *Mesopotamia and the Bible.* Continuum International Publishing Group, New York.

Dunbar, R. I. M. (2020) Religion, the social brain and the mystical stance. *Archive for the Psychology of Religion.* 42(1):46-62.

Normandi Ellis' translation of the *Egyptian Book of the Dead,* published by Phanes Press in 1991, Grand Rapids, MI.

Freeman, T. Do Jews Believe in Heaven? An overview of fundamental Jewish beliefs concerning reward and the afterlife. Chabad.org. https://www.chabad.org/library/article_cdo/aid/4848230/jewish/Heaven-and-the-Afterlife.htm

Garland, R. (1985) *The Greek Way of Death.* Ithaca, N.Y.: Cornell University Press, NY.

George, A. R. (2003). *The Babylonian Gilgamesh Epic: Introduction, Critical Edition and Cuneiform Texts.* Oxford University Press, UK. *p. 70.*

Grant, M. and J. Hazel (1993) *Who's Who in Classical Mythology.* Dent, London.

Jarrett, E. (2019) What is Heaven? A Complete Guide. Christianity.com, 22 August 2019.

King, C. W. (2020) *The Ancient Roman Afterlife: Di Manes, Belief, and the Cult of the Dead.* University of Texas Press, Austin.

Miller, L. (2010) *Heaven. Our Enduring Fascination with the Afterlife.* Harper Collins, New York.

Schneider, T. (2011). *An Introduction to Ancient Mesopotamian Religion.* William B. Eerdmans Publishing Company, Grand Rapids, MI.

Chapter 6 Reincarnation Beliefs

Betts, R. B. (1990) *The Druze.* Yale University Press, New Haven, CT.

Cole, O. W. and P.S. Sambhi (2005) *A Popular Dictionary of Sikhism: Sikh Religion and Philosophy.* Routledge, Abington-on-Thames, England.

Cort, J. E. (2001) *Jains in the World: Religious Values and Ideology in India.* Oxford University Press, UK.

Harvey, P. (2012) *An Introduction to Buddhism: Teachings, History, and Practices.* Cambridge University Press.

Keown, D. (2013) *Buddhism: A Very Short Introduction,* Oxford University Press, UK.

Lopez Jr., D. S. (2020) *Asina Religions in Practice.* Princeton University Press, NY.

Mararim, S. N. (1974) *The Druze Faith.* Caravan Books, Mattan, Pakistan.

Mandair, A. S. (2013) *Sikhism: A Guide for the Perplexed.* Bloomsbury Academic, NY.

Nator M. and A. Shoshana. The Phenomenology of 'Solved' Reincarnation Stories Among Druze in Israel: Private Self, Symbolic Type and Daily Life. *Cult Med Psychiatry.* 2022 Jun;46(2):194-211

Nator, M. and A. Shoshana (2025) “Open” reincarnation stories: The dialectic between doubt and certainty. *Transcultural Psychiatry*, 62(4).

Scholem, G. (1962) *The Origins of the Kabbalah.* Schocken Books, New York.

Swayd, S. (2006) *Historical Dictionary of the Druzes.* Scarecrow Press, Toronto.

Takhar, O. K. (2016) *Sikh Identity: An Exploration of Groups Among Sikhs* Routledge, New York.

Dalai Lama homepage Https://www.dalailama.com

Chapter 7 Spiritualism, Mediums and the Afterlife

Alvarado, C. (2014) Visions of the Dying, by James H. Hyslop (1907). *History of Psychiatry* 25:237-252.

Anderson, R. (1885) The Life and Work of James H. Hyslop. *The Journal of the American Society for Psychical Research,* 79:167-204.

Beischel, J. and G. Schwartz (2007) Anomalous information reception by research mediums demonstrated using a novel triple-blind protocol. *Explore* 3(1):23-27.

Conan Doyle, A. (2009) *History of Spiritualism.* Cambridge Publishing Classics Texts. CSP Classic Texts, New Edition.

Foy, R. P. (2008) *Witnessing the Impossible.* Torcal Publications, Diss, Norfolk, U.K.

Lewis, E. E. (1848) *A Report of the Mysterious Noises Heard in the House of Mr. John D. Fox in Hydesville, Arcadia, Wayne County.* Authenticated by the Certificates and Confirmed by Statements of the Citizens of that Place and Vicinity. Power Press of Shepard and Reed, Rochester, NY.

Owen, G. V. (1921) *The Life Beyond the Vale: A Compilation of Four Classic Books.* Republished by Square Circles Publishing, Pahrump, NV 2014.

Owen, G. V. (1921) *The Outlands of Heaven: The Life Beyond the Veil Volume 5.* Reprinted by lulu.com 2017

Pleasants, H. (1964). *Biographical Dictionary of Parapsychology.* Garrett Publications, Australia.

Melton, J. G. (2001) editor. "Hyslop, James Hervey." *Encyclopedia of Occultism & Parapsychology.* Vol 1. Gate Research Inc.

Stead, W. T., et al. (1922) *The Blue Island.* Self-published in London.

Chapter 8 Do Animals Go to Heaven?

Appleby, R., et al. (2013) Observations of a free-ranging adult female dingo (Canis dingo) and littermates' responses to the death of a pup. *Behavioural Processes,* 96:42-46

Bearzi, G., et al. (2018) Whale and dolphin behavioural responses to dead conspecifics. *Zoology,* 128:1-15.

Bercovich, F. (2013) Giraffe cow reaction to the death of her newborn calf. *African Journal of Ecology,* 51 (2):1-4.

Boesch, C. and H. Boesch-Ackermann (2000) *The Chimpanzees of the Tai Forest: Behavioral Ecology and Evolution.* Oxford: University Press, UK.

Broom, D. M., et al. (2009) Pigs learn what a mirror image represents and use it to obtain information. *Animal Behaviour,* 78(5):1037-1041.

Cazzolla Gatti, R. (2015). "Self-consciousness: beyond the looking-glass and what dogs found there". *Ethology Ecology & Evolution.* 28 (2):232–240.

De Kort, D., et al. (2017) Collared peccary (*Pecari tajacu*) behavioral reactions toward a dead member of the herd. *Ethology,* 134(2):131-134.

Douglas-Hamilton, I., et al. (2006) Behavioural reactions of elephants towards a dying and deceased matriarch. *Applied Animal Behaviour,*100 (1-2):87-102

Dudzonski, K. M., et al. (2003) Behavioural observations of bottlenose dolphins towards dead conspecifics. *Aquatic Mammals,* 29 (1):108-116.

Goldenberg, S. Z. and G. Wittemyer (2020) Elephant behavior toward the dead: A review and insights from field observations. *Primates,* 61(1):119-128.

Goodall, J. (1986) *The Chimpanzees of Gombe: Patterns of Behavior.* The Belknap Press of Harvard University Press, Cambridge, MA.

Herbst, C. T., et al. (2012) How Low Can You Go? Physical Production Mechanism of Elephant Infrasonic Vocalizations. *Science,* 337 (6094):595.

Iglesias, T. L., et al. (2012) Western scrub-jay funerals: cacophonous aggregations in response to dead conspecifics. *Animal Behaviour,* 84 (5):1103-1111.

Kaswan, P. and A. Roy. (2024) Unearthing calf burials among Asian Elephants Elephas maximus Linnaeus, 1758 (Mammalia: Proboscidea: Elephantidae) in northern Bengal, India. *Journal of Threatened Taxa,* 26 Tamil Nadu, India.

Martin, M. (2004) Elephant Destiny: Biography of an Endangered Species in Africa. *Canada: PublicAffairs.* pp. 184–186.

McComb, K., et al. (2014) Elephants can determine ethnicity, gender, and age from acoustic cues in human voices. *Proceeds of the National Academy of Sciences, USA* 111:5433–5438.

Pepperberg, I. M. (2002) Cognitive and communicative abilities of Grey Parrots. *Current Directions in Psychological Science,* 11(3):83-87.

Prior, H., et al. (2008). De Waal, Frans (ed.) Mirror-induced behavior in the Magpie (Pica pica): Evidence of self-recognition. *PLOS Biology,* 6(8).

Reggente, M. A. L., et al. (2016) Nurturant behavior toward dead conspecifics in free-ranging mammals: new records for odontocetes and a general review *Journal of Mammalogy,* 97(5):1428–1434.

Swift, K. N. and J. M. Marzluff (2015) Wild American crows gather around their dead to learn about danger. *Animal Behaviour,* 109:187-197.

Chapter 9 Upright Explorers

Anton, S. C. (2003) Natural History of *Homo erectus. American Journal of Biological Anthropology,* 537(122):126-170.

Bednarik, R. G. (2003) A figurine from the African Acheulian, Reports. *Current Anthropology,* 44(3):405-413.

Bednarik, R. G. (1993) Paleolithic art in India. *Man and the Environment,* 18 (2):33-40.

Bednarik, R. G. (1988) Comment on D. Mania and U. Mania 'Deliberate engravings on bone artefacts of Homo erectus' *Rock Art Research* 5:96-100.

Bednarik, R. G., et al. (2005) Preliminary Results of the EIP Project. *Rock Art Research,* 22(2):147-197.

Boaz, N. T. and R. Ciochon (2004) *Dragon Bone Hill: An Ice-Age Saga of Homo erectus.* Oxford University Press, UK.

Joordens, J. C., et al. (2015) *Homo erectus* at Trinil on Java used shells for tool production and engraving. *Nature,* 518:228-231.

Kumar, G. and R. Krishna. (2014) Understanding the Technology of the Daraki-Chattan Cupules. *Rock Art Research,* 31(2):177-186.

Mania, D. (1995) The earliest occupation of Europe: the Elbe- Saale region (Germany). In: W. Roebroeks/T. van Kolfschoten (eds.) *The Earliest Occupation of Europe.* Analecta Leidensia (Leiden), 85-101.

Mania, D. (1995) Bilzingsleben - middle Pleistocene site of *Homo erectus.* Travertine complex and fauna at Bilzingsleben. *Quaternary field trips in Central Europe,* 14. Congress INQUA (Berlin), 738-740, 777-780, 1078-1079.

Pettitt, P. (2011) *The Paleolithic Origins of Human Burial.* Routledge, New York.

Chapter 10 Archaic Peoples

Bednarik, R. G. (2014) Pleistocene Paleoart of Europe. *Arts,* 3:245-278.

d'Errico, F., et al. (2001) An Engraved Bone Fragment From c. 70,000-Year-Old Middle Stone Age Levels at Blombos Cave, South Africa: Implications for the Origin of Symbolism and Language. *Antiquity,* 75:309–318.

Henshilwood, C. S., et al. (2011) A 100,000-Year-Old Ochre-Processing Workshop at Blombos Cave, South Africa. *Science,* 334:219–222.

Mackay, A. and A. Welz (2008) Engraved ochre from a Middle Stone Age context at Klein Kliphuis in the Western Cape of South Africa. *Journal of Archaeological Science,* 35:1521–1532.

Marean, C., et al. (2007) Early human use of marine resources and pigment in South Africa during the middle Pleistocene. Nature, 449(7164)905-908.

Sarmiento, E. and M. Pickford (2022) Muddying the muddle in the middle even more. *Evolutionary Anthropology,* 31 (5):237-239.

Stringer, C. B. (2012) The status of *Homo heidelbergensis* (Schoetensack 1908) *Evolutionary Anthropology,* 21 (3):101-104.

Stringer, C. (1993) Secrets of the Pit of the Bones. *Nature,* 362:501–502

Villa, P. (1983) *Terra Amata and the Middle Pleistocene archaeological record of southern France.* University of California Press, Berkeley, CA.

Chapter 11 People of the Ice Ages

Balzeau, A., et al. (2020) Pluridisciplinary evidence for burial for the La Ferrassie 8 Neandertal child. *Scientific Reports,* 10:21230.

Been, E. and O. Barzilai (2024) Neandertal burial practices in Western Asia: How different are they from those of the early Homo sapiens? *L'Anthropologie,* (103281).

Bello, S. M. (2021) Boning up on Neanderthal art. *Nature Ecology & Evolution,* 5:1201–1202

Frayer, D. W., et al. (2006) Krapina 3: Cut marks and ritual behavior? *Periodicum Biologorum,* 108(4):519-524.

Hoffmann, D.L., et al. (2018) U-TH dating of carbonate crusts reveals Neanderthal origin of Iberian cave art. *Science,* 359(6378):912-915.

Jaubert, J., et al. (2016) Early Neanderthal constructions deep in Bruniquel Cave in southwestern France. *Nature,* 534:11-114

Leroi-Gourhan, A. (1975) The flowers found with Shanidar IV, a Neanderthal burial in Iraq. *Science,* 190:562-64.

King, W. (1864) The reputed fossil man of the Neanderthal. *Quarterly Review of Science,* 1:88-97.

Pomeroy, E., et al. (2020) New Neanderthal Remains Associated with the 'Flower Burial' at Shanidar Cave. *Antiquity,* 94(373):11-26.

Radovčić D., et al. (2015) Evidence for Neandertal Jewelry: Modified White-Tailed Eagle claws at Krapina. *PLoS ONE,* 10 (3):e0119802

Rendu, W., et al. (2013) Evidence supporting an intentional Neandertal burial at La Chapelle-aux-Saints, *Anthropology,* 111 (1):81-86

Roebroeks, W., et al. (2012) Use of red ochre by early Neanderthals. *Proceedings of the National Academy of Sciences,* 109(6):1889-1894.

Solecki, R. S. (1975). Shanidar IV, a Neanderthal Flower Burial in Northern Iraq. *Science,* 190 (4217):880–881.

Trinkhaus, E. (1985) Cannibalism and burial at Krapina. *Journal of Human Evolution,* 14(2):203-216.

Trinkhaus, E. (1985) Pathology and the posture of the La Chappelle-aux-Saints Neanderthal. *American Journal of Physical Anthropology,* 67:19-41.

Trinkaus, E. and P. Shipman (1993) *The Neanderthals: Changing the Image of Mankind.* Knopf, New York.

Zilhao, J., et al. (2010) Symbolic use of marine shells and mineral pigments by Iberian Neanderthals. *Anthropology,* 107 (3):1023-1028.

Chapter 12 The Wise Humans

Aubert, M., et al. (2014) Pleistocene cave art from Sulawesi, Indonesia. *Nature*, 514:223–227.

Bacon, B., et al. (2023) An Upper Palaeolithic Proto-writing System and Phenological Calendar. *Cambridge Archaeological Journal*, 33(3):1-19.

Bar-Yosef Mayer, D. E., et al. (2009) Shells and ochre in Middle Paleolithic Qafzeh Cave, Israel: indications for modern behavior. *Journal of Human Evolution* 56(3):307-314.

Bednarik, R. G. (2001) The oldest known rock art in the world. *Anthropology*, 39(2/3):89-98.

Benson, L. V., et al. (2013) Dating North America's oldest petroglyphs, Winnemucca Lake subbasin, Nevada. *Journal of Archaeological Science*, 40(12):4466-4476.

Brumm, A., et al. (2021) Oldest Cave art found in Sulawesi. *Science Advances*, 7:eabd4648.

Bowler, J. M., et al. (2003). New ages for human occupation and climatic change at Lake Mungo, Australia. *Nature* 421:837-840.

Durband, A. C. R., et al. (2009). A new test of the sex of the Lake Mungo 3 skeleton. *Archaeology in Oceania* 44(2):77-83.

Gargett, R. H. (1999) Middle Paleolithic burial in not a dead issue: the view from Qafzeh, Saint-Cesaire, Kebara, Amud and Dederiyeh. *Journal of Human Evolution*, 37(1):27-90.

Hovers, E., et al. (2003). An early case of color symbolism: Ochre use by modern humans in Qafzeh Cave. *Current Anthropology*, 44(4):491-522.

Lanese, N. (2021) Kids fossilized handprints may be some of the world's oldest art. *Scientific American*, 17 October 2021.

Neves, A. G. M. and M. Serva (2012) Extremely rare interbreeding events can explain Neanderthal DNA in living humans. *PLoS ONE*, 7(10): e47076

St. Fleur, N. (2018) Oldest known drawings by human hands discovered in South African cave. *The New York Times*. 15 September 2018.

Trinkaus, E. and A. P. Buzhilova (2018) Diversity and differential disposal of the dead at Sunghir. *Antiquity*, 92(361):7-21.

Whitley, D. S. (2009) *Cave Paintings and the Human Spirit: The Origin of Creativity and Belief.* Prometheus Books, Amherst, MA.

Chapter 13 Megaliths

Baiju, C. K., et al. (2024) Environmental forces shaping settlement and subsistence of ancient civilization: Tracing the patterns in Marayoor, Western Ghats, India. *Journal of Geointerface* 3(1):44-55.

Gresky, J., et al. (2017) Modified human crania from Gobekli Tepe provide evidence for a new form of Neolithic skull cult. *Scientific Advances*, 3(6):e1700564.

Hodder, I. R. (Ed.) (2013) *Humans and landscapes of Çatalhöyük: reports from the 2000-2008 seasons.* Çatalhöyük Reseach Project Series Volume 8. British Institute at Ankara Monograph No. 47/ Monumenta Archaeologica 30, Los Angeles: Cotsen Institute of Archaeology Press.

Hodder, I.R. (Ed.) (2013) *Substantive technologies at Çatalhöyük: reports from the 2000-2008 seasons.* Çatalhöyük Reseach Project Series Volume 9, British Institute at Ankara Monograph No. 48 / Monumenta Archaeologica 31, Los Angeles: Cotsen Institute of Archaeology Press.

Hodder, I.R. (Ed.) (2014) *Çatalhöyük excavations: the 2000-2008 seasons.* Çatalhöyük Reseach Project Series Volume 7, British Institute at Ankara Monograph No. 46 / Monumenta Archaeologica 29, Los Angeles: Cotsen Institute of Archaeology Press.

Hodder, I.R. (Ed.) (2014) *Integrating Çatalhöyük: themes from the 2000-2008 seasons.* Çatalhöyük Reseach Project Series Volume 10, British Institute at Ankara Monograph No. 49 / Monumenta Archaeologica 32, Los Angeles: Cotsen Institute of Archaeology Press.

Pearson, M. P. (2012) *Stonehenge.* Simon and Schuster, Ltd., UK

Peters, J. and K. Schmidt (2004) Animals in the symbolic world of pre-pottery Neolithic Goblekli Tepe, south-eastern Turkey: a preliminary assessment. *Anthropozoologica,* 39(1):179-218.

Sweatman, M. B. (2021) The Younger Dryas impact hypothesis: Review of the impact evidence. *Earth Science Reviews,* 218:103677.

Sweatman, M. B. and D. Tsikritsis (2017) Decoding Golgekli Tepe with archaeoastronomy: What does the fox say? *Mediterranean Archeology and Archaeometry,* 17(1):233-250.

(no author) Crucial new data on the origin of the Dolmens of Antequera, a World Heritage Site Popular archaeology. June 28, 2018

Antequera Dolmens Site, UNESCO, World Heritage Convention https://whc.unesco.org/en/list/1501/

https://worldheritagesites.net/antequera-dolmens/

Newgrange World Heritage Site https://newgrange.com

Dolmens of Marayoor https://enrouteindianhistory.com/the-megalithic-dolmens-of-marayoor/

Chapter 14 Consciousness: The Philosophers

Barušs, I. (2016) "Consciousness: Where We Are At." In *CONSCIOUSNESS: Ideas and Research for the Twenty-First Century,* 5(5): Article 1.

Goff, P. (2019) *Galileo's Error: Foundations for a New Science of Consciousness.* Vintage Press, New York.

Goff, P. (2017) The Case for Panpsychism. *Philosophy Now.* Issue 121.

Koch, C., et al. (2016) Neural correlates of consciousness: progress and problems. Nature Reviews/*Neuroscience,* 17:307-321.

Overgaard, M. (2017) The Status and Future of Consciousness. *Research Frontiers Psychology, 8.*

Sullivan, P. R. (1995). Contentless consciousness and information-processing theories of mind. *Philosophy, Psychiatry, & Psychology,* 2(1):51–59.

Tononi, G. and C. Koch (2008) The neural correlates of consciousness: an update. *Annual New York Academy Sciences,* 1124:239–261.

Tononi, G. and C. Koch (2015) Consciousness: Here, There and Everywhere? *Philosophical Transactions B,* 370(1668).

Tononi. G., at al. (2016) Integrated information Theory: From consciousness to its physical substrate. *Nature Reviews Neuroscience,* 17:450-461.

Vibert, D. A. (1956) *The Life of Arthur Eddington.* Thomas Nelson and Sons, Edinburgh, Scotland. pp.92-95

The Bertrand Russell Society http://bertrandrussellsociety.org

Chapter 15 Consciousness: The Neuroscientists

Brefczynski-Lewis, J. A., et al. (2007) Neural correlates of attentional expertise in long-term meditation practitioners. *Proceedings of the National Academy of Sciences* 107(27):11483-11488.

Davidson, R. J. (2006) Science and Spirituality: Probing the convergences and tensions. Review of the universe in a single atom: The convergence of science and spirituality by His Holiness the Dalai Lama. PsycCRITIQUES, 51(28).

Davidson, R. J. and C. J. Dahl (2017) Outstanding challenges in scientific research on mindfulness and meditation. *Perspectives on Psychological Science*, 13(1):62-65.

De Graaf, T. A., et al. (2012) The 'correlates' in neural correlates of consciousness. *Neuroscience & Biobehavioral Reviews*, 36(1):191-197.

Demertzi, A., et al. (2019) Human consciousness is supported by dynamic complex patterns of brain signal coordination *Science Advances*, 5(2).

Ferrarelli, F., et al. (2013) Experimental mindfulness meditators exhibit higher parietal-occipital EEG gamma activity during NREM sleep. *PLoS ONE*, 8(8): e73417.

Josipovic, Zoran, et al. (2012) Influence of meditation on anti-correlated networks in the brain, *Frontiers of Human Neuroscience*, Jan 2012.

Koch, Christof (2018) "What Is Consciousness?" *Scientific American* 318(6):60-64.

Kozhevnikov, M., et al. (2009). The Enhancement of Visual- Spatial Processing Efficiency through Buddhist Deity Meditation. *Psychological Science*, 20(5):645-653.

Merker, B. (2007) Consciousness without a Cerebral Cortex: A Challenge for Neuroscience and Medicine. *Behavioral and Brain Sciences*,30:63–81.

Phillippi, C. L., et al. (2012) Preserved Self-Awareness following Extensive Bilateral Brain Damage to the Insula, Anterior Cingulate, and Medial Prefrontal Cortices. PLoS ONE, 7(8).

Yang C. C., et al. (2019) Alterations in brain structure and amplitude of low-frequency after 8 weeks of mindfulness meditation training in meditation-naïve subjects. *Scientific Reports*, 9:10977.

Chapter 16 Near-Death Experiences

Charlier, P. (2014) Oldest medical description of a near death experience (NDE), France, 18th century. *Resuscitation*, 85(9).

De Benneville, G. (1804). The remarkable account of the life and trance of Doctor George de Benneville. Philadelphia, PA: Thomas T. Stiles. (Available as U.S. Government Microform)

Plato (1892). *The Dialogues of Plato* (Jowett, B., trans.). Humphrey Milford, London. (Original work published 4th century B.C.

Plutarch. *Plutarch's Morals.* Translated from the Greek by several hands. Corrected and revised by. William W. Goodwin, PH. D. Little, Brown, and Company, Cambridge. Press Of John Wilson and son. 1874.

Vincent, K. R. and J. C. Morgan (2006) An 18th century Near Death Experience: The Case of George de Benneville. *Journal of Near-Death Studies,* 25(1):35-48.

Zaleski, C. (1987) *Otherworld Journeys: Accounts of Near-Death Experience in Medieval and Modern Times.* Oxford University Press, New York.

Chapter 17 Seers

Cayce, E. (2008) *Beyond Death: Visions from the Other Side.* A.R.E Press, Virginia Beach, VA.

Davis, A. J. (1857) *The Magic Staff: An Autobiography of Andrew Jackson Davis.* Reprinted 2021

Davis, A. J. (1865) *Death and the Afterlife.* Reprinted 2015 by Square Circles Publishing, Schenectady, NY.

Robinson, J. M. (1990) *The Nag Hammadi Library.* HarperOne, New York.

Sawtrey (H. of Saltrey) (1180-1184) *The Treatise on the Purgatory of St. Patrick.*

Seymour, J. D. (2022) *St Patrick's Purgatory: A Mediaeval Pilgrimage in Ireland.* Legare Street Press, NY.

Swedenborg, E. and L. Donald (2006) *Afterlife: A Guided Tour of Heaven and Its Wonders, Second Edition.* Swedenborg Foundation Publishers, West Chester, PA.

Chapter 18 Near-Death Experiences

Alexander, E. (2012) *Proof of Heaven: A Neurosurgeon's Journey into the Afterlife.* Simon and Schuster

Greyson, B. and I. Stevenson (1980) The phenomenology of near-death experiences. *American Journal of Psychiatry,* 137:1193-1196.

Greyson, B. (1983) The near-death experience scale: Construction, reliability, and validity, *The Journal of Nervous and Mental Disease*, 171(6):369-375.

Greyson, B. (2006) Near-death experiences and spirituality. *Zygon,* 41:393-414.

Kirkley, D. D. (1975) *The Psychology of Blindness.* Nelson-Hall, Chicago

Moody, R. A. (2013) Getting comfortable with death & near-death experiences: Near-death experiences: An essay in medicine & philosophy. *Missouri Medicine,*110(5):368-371.

Moody, R. A. (1977) *Reflections on Life After Life,* Stackpole Books, Harrisburg, PA.

Ring, K. and S. Cooper (1997) Near-death and out-of-body experiences in the blind: A study of apparent eyeless vision. *Journal of Near Death Studies* 16(2):101-147.

Chapter 19 Out of Body Experiences

Carter, C. (2011) Response to "Could Pam Reynolds hear?" *Journal of Near-Death Studies, 30*(1):29–53.

Greyson, B. and D. James, eds. (2009) *The Handbook of Near-Death Experiences Thirty Years of Investigation.* Praeger Publishers, Santa Barbara, CA.

Sharp, K. C. (1995) *After the Light: What I discovered on the other side of life that can change your world.* William Morrow, New York.

Sharp, K. C. (2007) The other shoe drops: commentary on "Does paranormal perception occur in near-death experiences?" *Journal of Near-Death Studies,* 25(4):245.

Woerlee, G. M. (2011) Could Pam Reynolds hear? A new investigation into the possibility of hearing during this famous near-death experience. *Journal of Near-Death Studies, 30*(1):3–25.

Chapter 20 Ghost Hunting

Baruss, I. (2001) Failure to Replicate Electronic Voice Phenomenon. *Journal of Scientific Exploration,* 15(3):355.

Baruss, I. (2023) *Death as an Altered State of Consciousness: A Scientific Approach.* American Psychological Association, 2023.

Black, J. (1922) The spirit-photograph fraud: The evidence of trickery, and a demonstration of the tricks employed. *Scientific American,* 127:224-225.

George, S. C. (2021) Ghost Busted: When Science Meets Paranormal Activity. *Discover Magazine,* 7 Oct 2021

Kaplan, L. (2008) *The Strange Case of William Mumler, Spirit Photographer.* University of Minnesota Press, Minneapolis.

Permutt, C. (1983) *Beyond the Spectrum: Survey of Supernormal Photography.* Patrick Stephens Publishers Ltd., UK.

Potts, J. (2004). Ghost hunting in the twenty-first century. In J. Houran (Ed.), *From Shaman to scientist: essays on humanity's search for spirits* (pp. 211-232). Scarecrow Press, Lanham, MD.

Prince, W. F. (1925) My doubts about spirit photographs. *Scientific American,* 133:370–371.

Raudive, K. (1968) *Breakthrough: An Amazing Experiment in Electronic Communication with the Dead* in 1968 (Translated by N. Fowler, ed. By J. Morton into English in 1971).

Von Szalay, A. and R. Bayless (1959) Letter to the Editor January 1959. *American Society for Psychical Research.*

Warren, R. M. (1968) Verbal transformation effect and auditory perceptual mechanisms. *Psychological Bulletin,* 70 (4):261-270.

Chapter 21 Testing Mediums

Beischel, J. and G. Schwartz (2007) Anomalous information reception by research mediums demonstrated using a novel triple-blind protocol. *Explore* 3(1):23-27.

Foy, R. P. (2008) *Witnessing the Impossible.* Torcal Publications, Diss, Norfolk, U.K.

https://www.thescoleexperiment.com/trans-dimensional-communication-experiments.html

Chapter 22 Reincarnation Investigations

Keil, H. H. and J. B. Tucker (2000) An unusual birthmark case thought to be linked to a person who had previously died. *Psychological Reports,* 87:1067-1074.

Loftus, E. F. and J. E. Pickrell (1995). The formation of false memories. *Psychiatric Annals, 25:*720-725.

Stevenson, I. (1997) *Reincarnation and Biology Volume 1: Birthmarks.* Praeger, Westport, CT.

Tucker, J. B. (2016) The Case of James Leininger: An American Case of the Reincarnation Type. *Explore,* 12:200-207.

Tucker, J. B. (2013) *Return to Life: Extraordinary Cases of Children Who Remember Past Lives.* St. Martin's Griffin, New York.

Tucker, J. B. (2001) Measuring the strength of children's claims to remember previous lives. In *Trends in Rebirth Research: Proceedings of an International Seminar,* Senanayake N, (ed.). Ratmalana, Sri Lanka: Sarvodaya Vishva Lekha.

Tucker, J.B. (2000) A scale to measure the strength of children's claims of previous lives: methodology and initial findings. *Journal of Scientific Exploration,* 14(4):571-581.

Chapter 23 OBE Observations

Alvarado, C. S. (2000). Out-of-body experiences. In E. Cardeña, S. J. Lynn, & S. Krippner (Eds.), *Varieties of anomalous experience: Examining the scientific evidence* (pp. 183–218). American Psychological Association.

Moix, J., et al. (2025) Out of Body experiences: Scoping review. *Explore* 21(4).

Osis, K. (1978) Out-of-body Research at the American Society for Psychical Research. *In Mind Beyond the Body: The Mystery of ESP Projection, ed. by D. Scott Rogo.* Penguin Books, New York.

Osis, K. and D. McCormick (1980) Kinetic effects at the ostensible location of an out-of-body projection during perception testing. *Journal of the American Society for Psychical Research,* 74:319-29.

Sabom, M. B. (1981) The near-death experience: Myth or reality? A methodological approach. *Anabiosis: The Journal of Near-Death Studies, 1*(1):44–56.

Sabom, M. B. (1982) *Recollections of Death: A Medical Investigation.* Harper Collins, New York.

Chapter 24 Related Experiences

Chi-Yeon, L., at al. (2020) Terminal lucidity in the teaching hospital setting. *Death Studies* 44(5):285-291.

Depner, R. M., et al. (2020) Expanding the understanding of end-of-life dreams and visions: A consensual qualitative research analysis. *Palliative Medicine Reports.* 1(1):103-110.

Griffiths R. R., et al. (2019) Survey of subjective "God encounter experiences": Comparisons among naturally occurring experiences and those occasioned by the classic psychedelics psilocybin, LSD, ayahuasca, or DMT. *PLoS ONE,* 14(4): e0214377.

Martial, C., et al. (2019) Neurochemical models of near-death experiences: A large-scale study based on the semantic similarity of written reports. *Consciousness and Cognition,* 69:52-69.

Moody, R. A. and P. Perry (2010). *Glimpses of Eternity.* Guideposts.

Nahm, M. and B. Greyson, (2014) The Death of Anna Katharina Ehmer: A Case Study in Terminal Lucidity. *Journal of Death and Dying,* 68(1):77-87.

Nahm, M. and B. Greyson (2009). Terminal lucidity in patients with chronic schizophrenia and dementia: A Survey of the Literature. *Journal of Nervous and Mental Disease,* 197:942-44.

Nahm, M., et al. (2012). Terminal lucidity: A review and a case collection. *Archives of Gerontology and Geriatrics,* 55(1):138-142.

Nahm, M. (2011) Reflections on the context of near-death experiences. *Journal of Scientific Exploration,* 25:453-78.

Nahm, M. (2009) Terminal lucidity in people with mental illness and other mental disability: An overview and implications for possible explanatory models. *Journal of Near-Death Studies,* 28:87-106.

Shared crossing Research Initiative (SCRI). (2021) Shared Death Experiences: A Little-Known Type of End-of-Life Phenomena Reported by Caregivers and Loved Ones. *American Journal Hospice Palliative Care,* 12:1479-1487

Sweeney M. M., et al. (2022) Comparison of psychedelic and near-death or other non-ordinary experiences in changing attitudes about death and dying. *PLoS ONE,* 17(8):e0271926.

Chapter 25 OBEs, NDEs and the Brain

Blackmore, S. J. (1996) Near-Death Experiences. *Journal of the Royal Society of Medicine* 89(2):73-76.

Blanke, O., et al. (2002) Stimulating illusory own-body perceptions. *Nature,* 419(6904):269-70.

Blanke, O. and S. Arzy, (2005) The out-of-body experience: disturbed self-processing at the temporo-parietal junction. *Neuroscientist* 11:16-24.

Blanke, O. and C. Mohr, C. (2005) Out-of-body experience, heautoscopy, and autoscopic hallucination of neurological origin: Implications for neurocognitive mechanisms of corporeal awareness and self-consciousness. *Brain Research Reviews* 50:184-99.

Blanke, O., et al. (2016) Leaving body and life behind: Out-of-body and near-death experience. In Laureys, S, Gosseries, O., & Tononi, G. (Eds.) *The Neurology of Consciousness (Second Edition)*, Academic Press, Cambridge, MA.

Borjigin, J., et al (2013) Surge of neurophysiological coherence and connectivity in the dying brain. *Proceeds of the National Academy of Sciences USA*, 110(35):14432-7.

Braithwaite, J.J., et al. (2011) Cognitive correlates of the spontaneous out-of-body experience (OBE) in the psychologically normal population: evidence for an increased role of temporal-lobe instability, body-distortion processing, and impairments in own-body transformations *Cortex*, 47(7):839-53.

Chawla, L.S., et al. (2017) Characterization of end-of-life electroencephalographic surges in critically ill patients. *Death Studies* 41:385-392.

Greyson, B., et al. (2014) Out-of-body experiences associated with seizures. *Frontiers of Human Neuroscience*, 8:65.

Holden, J. M., et al. (2009). The field of near-death studies: Past, present, and future. In J. M. Holden, B. Greyson, & D. James (Eds.), *The handbook of near-death experiences: Thirty years of investigation* (pp. 1–16). Praeger/ABC-CLIO.

Lopez, C. and M. Elziere (2018) Out-of-body experience in vestibular disorders - A prospective study of 210 patients with dizziness. *Cortex*, 104:193-206.

Martial, C., et al. (2020) Near-death experience as a probe to explore (disconnected) consciousness. *Trends in Cognitive Sciences*, 24:173-183

Neppe, V. (2011) Models of the out-of-body experience: A new multi-etiological phenomenological approach. *NeuroQuantology Journal* 9(1).

Smith, S. A. and C. Messier (2014) Voluntary out-of-body experience: an fMRI study. *Frontiers of Human Neuroscience*, 8:70.

Timmermann, C., *et al.* (2018) DMT models the near-death experience. *Frontiers in Psychology*, 9:1-12.

Van Gordon, W., et al. (2018) Meditation-induced near-death experiences: a 3-year longitudinal study. *Mindfulness*; 9(6):1794-1806.

Vincente, R., et al. (2022) Enhanced interplay of neuronal coherence and coupling in the dying human brain. *Aging Neuroscience, 14.*

Chapter 26 Is Testing Possible?

Evrard, R., et al. (2022) Sawing the branch of near-death experience research: A critical analysis of Parnia *et al.*'s paper. *Annals of the New York Academy of Sciences,* 1515:5-9.

Parnia, S. and J. Young (2013) *Erasing Death: The Science That Is Rewriting the Boundaries Between Life and Death.* HarperOne, New York.

Parnia, S. (2007) Do reports of consciousness during cardiac arrest hold the key to discovering the nature of consciousness? *Medical Hypotheses,* 69 (4):933.

Parnia, S. (2014) Death and consciousness—an overview of the mental and cognitive experience of death. *Annals of the New York Academy of Sciences,*1330:75-93.

Parnia, S., at al. (2014) AWARE–AWAreness during REsusitation–A prospective study. *Resuscitation,* 85(12): 1799-1805.

Parnia, S., et al. (2022) Guidelines and standards for the study of death and recalled experiences of death–a multidisciplinary consensus statement and proposed future directions. *Annals of the New York Academy of Sciences,* 1511:5-21.

Sabom. M. (1998) *Light and Death: One doctors fascinating account of near-death experiences.* Zondervan, Grand Rapids, MI

Woerlee, G.M. (2011). Could Pam Reynolds hear? A new investigation into the possibility of hearing during this famous near-death experience. *Journal of Near-Death Studies,* 30(1):3-25.

Sabom, M. (2007). Commentary on "Does paranormal perception occur in near-death experiences? *Journal of Near-Death Studies,* 25(4):257-60.

Xu, G., et al. (2023) Surge of neurophysiological coupling and connectivity of gamma oscillations in the dying human brain. *Proceedings of the National Academy of Sciences, USA,* 120(19): e2216268120

.

About the Author

Susan Urbanek Linville earned a PhD in biology from the University of Dayton and has experience as an assistant editor for a science journal, university outreach coordinator and museum assistant administrator. As a freelance writer, she has published short fiction, newspaper and magazine articles, non-fiction books, blog articles, and was a script writer for Indiana University's *A Moment of Science* Podcast Series. Her non-fiction books include an African memoir, Pennsylvania Underground Railroad history, and several books on Spiritualism.

Made in the USA
Monee, IL
04 November 2026

67383948R00148